GLOBAL SCIENCE LITERACY

Science & Technology Education Library

VOLUME 15

SCOPE
The book series *Science & Technology Education Library* provides a publication forum for scholarship in science and technology education. It aims to publish innovative books which are at the forefront of the field. Monographs as well as collections of papers will be published.

The titles published in this series are listed at the end of this volume.

Global Science Literacy

Edited by

VICTOR J. MAYER

*The Ohio State University,
Columbus, U.S.A.*

KLUWER ACADEMIC PUBLISHERS
DORDRECHT / BOSTON / LONDON

A C.I.P. Ctalogue record for this book is available from the Library of Congress.

ISBN 1-4020-0514-8

Published by Kluwer Academic Publishers,
P.O. Box 17, 3300 AA Dordrecht, The Netherlands.

Sold and distributed in North, Central and South America
by Kluwer Academic Publishers,
101 Philip Drive, Norwell, MA 02061, U.S.A.

In all other countries, sold and distributed
by Kluwer Academic Publishers,
P.O. Box 322, 3300 AH Dordrecht, The Netherlands.

Printed on acid-free paper

To My Asian Friends and Associates

I have had the good fortune to be a teacher, a colleague and a friend of many Asian science educators, especially from Korea, Japan, Taiwan and the People's Republic of China. Many of the opportunities to meet and work with these exceptional people were caused in some way by Dr. Kwon Jae-sool now of the Korea National University of Education. I first met him as a student. We became colleagues in science education.

Such close professional collaborations and friendships have enabled me to work for extended periods in Japan and Korea. The resulting in-depth cultural experiences have been crucial in developing many of the insights found in the chapters of this book. Those the reader might find valuable and insightful come from these long-term associations and cultural experiences. Those that are not valuable and insightful are the result of the adherence I may have to my Western upbringing.

It seems only appropriate then for me to dedicate this book on Global Science Literacy, not only to Dr Kwon, but to all of those Asian science educators that I have had the good fortune of working with these past 30 or so years.

CONTENTS

PREFACE

Earth is unique, a planet of rare beauty and great value.

I firmly believe that conclusions about global warming reached by researchers, such as those at my university in the Byrd Polar Research Center, are valid and based on very solid science. They are telling us that humans, by their actions, are causing the global climate to change. It therefore puzzles me when politicians, representatives of the business community, and even some physicists and chemists, make public statements that this is "junk" science. Or that it is just another "chicken little" scare of the "environmentalists". Studies on public understanding of science among Americans by Jon Miller and his colleagues and the current political power of those advocating "creation science" as a component of the science curriculum, suggests to me that there is a deep misunderstanding of the nature of science here in the United States, perhaps shared by some.scientists If the resources devoted to the improvement of science teaching and curriculum following World War II had been effectively used I would expect that there would be an acceptance of good science by influential members of our society, and by most citizens. However, I do believe that the programs supported by the National Science Foundation over the past forty years have been effective in establishing strong science programs in our schools. After all, my colleagues and I have been recipients of some of that support for curriculum and teacher enhancement programs.

My hypothesis concerning this apparent paradox is that these science teaching renewal programs have given priority to the perceived needs of a country in conflict. As a result, they focused on the type of science supporting our technical needs, neglecting an equally important part of science, what we in this book are calling the "system sciences". Such neglect has continued throughout the science restructuring programs taking place in the USA and elsewhere around the world-- thus, the need for this book.

The need to re-examine our goals for science curricula

We have just ended a long period of conflict between the major nations, the Cold War. Science and science education played a central role in waging that war and the "hot wars" that preceded it. Now, however, science education and its practitioners have the opportunity of supporting changes in the goals of science as it adjusts to a new era. Science is being challenged by some to provide the knowledge to counter the devastating environmental problems that have been by-products of a century of war and economic conflict. It also can be employed to help solve the social problems resulting from the unfettered use of technology for political and economic gain. To support science in redirecting its goals, science educators must re-examine the very nature of science and its role in social, cultural and political systems. We must understand the broad nature of science and its methodologies, an understanding not always apparent in the professional dialog of science educators. It is our belief that such a re-examination will result in a significant change in science education; a change founded on the view that science is, after all, a study of the Earth System in which we all live, not simply the basis for the pursuit of ever more technology.

This book offers a rationale and a developmental basis for such a re-examination. Authors from six countries, representing East and West, provide support and ideas for application of a different approach to the nature of the science curriculum. One we have called, Global Science Literacy. Most authors come from an Earth science academic background, but there also are physicists, chemists, and biologists represented as authors or co-authors of chapters. Global Science Literacy (GSL) is based on developments in the United States that resulted in an approach called Earth Systems Education (ESE), a curricular basis for literacy in science.

Earth Systems Education (ESE) uses the Earth system as the organizing conceptual theme for developing science curricula for the middle through high school levels. Children of all nations experience weather, flowing streams, and rock materials as parts of their environment. They observe the beauty of sunsets, the power of storms, the tranquility of a mountain scene, a flowing river, or an autumn day. A science curriculum organized around students' interdependence with nature and tapping into their interests in nature provides a common subject for study in all cultures. ESE includes the science methodology of the system sciences, a distinct contrast from the prevailing emphasis upon that of the physical sciences in the world's science curricula. A facility with the use of science methodology can provide the world's future citizens with universal methods of communication and problem solving as they enter the adult world.

This book is, in part, the result of an international process of expanding ESE into a Global Science Literacy program that includes a cooperative effort between faculties of The Ohio State University and Hyogo University of Teacher Education (Japan). A seven-month long global education project at Hyogo University provided an opportunity to synthesize many ideas that had evolved over the years of involvement by the authors in Earth science education. We have formulated a global version of science literacy for pre-college education that will not only improve citizens' understanding of science, but also enhance cross-cultural communication and understanding. We have also examined selected Asian cultures and drawn implications for a science program more in concert with those cultures, especially with its incorporation of system science methodology. In addition, a Fulbright grant for a subsequent project allowed us to evaluate Global Science Literacy as a basis for curriculum development at the upper secondary school level in Japan. This project was located at Shizuoka University and in the Division of Educational Research of Monbusho. Through contacts made during international conferences and professional trips, science educators from thirteen countries have become involved in enhancing and expanding ideas relating to Global Science Literacy.

The resulting matrix of science and social concepts and processes are proposed as a functional international definition of science literacy. If implemented in school curricula of democratic nations, we believe it will help citizens understand the role of science in solving environmental and social problems left in the wake of a century of world war and economic conflict. It can also contribute to cross-cultural understanding and cooperation between citizens and leaders of the democracies of the world. Thus, science curricula can have a crucial role among other curricular subjects in helping students achieve a global understanding and perspective--a major objective of the social studies curriculum construct of global education.

Organization of the book

In the first section of the book, we lay the historical, conceptual and philosophical groundwork for GSL, a science curriculum, international in scope, conceptual in organization, centered on the students and their habitats and representative of the very broad methodology used by scientists. In the second section, authors discuss a variety of learning environments, which, though not new to GSL, are supportive of the basic goals of the curriculum effort. These environments include the Internet, cooperative learning, effective reading materials, field investigations, and authentic assessment. In the third and last section, certain issues in curriculum development are discussed with a GSL perspective. How does one develop curriculum that is conceptually organized rather than organized by the traditional disciplines? How can GSL curricula be adapted for special learners? How can field activities and aesthetics be integrated in GSL curricula? The final chapter reports the research study conducted in Japan that looked at the feasibility of implementing GSL curricula at the upper secondary school level in that country.

Personal reflections

One of the best practitioners of "applied science" I knew was my father. I grew up on a farm in Wisconsin. My father went no further than eighth grade in his formal schooling. From when he was a child until he retired, he worked almost every day milking cows, planting and harvesting hay and grains, and the various other duties that went into being a successful farmer. Helping him as a child, I did not understand why he did certain things. In the spring, he would plow fields following the contours of the land instead of up and down hills even though that would often have been much easier. He also alternated different crops in parallel strips around the hills. In successive years, he would alternate crops within a single strip. He allowed trees and brush to grow along fencerows harboring animals and birds that often fed off the crops he was raising. Although he kept a bull on the farm for breeding the cows, after I became a teen-ager he sold it and joined an artificial breeding cooperative. He soon became a member of its board of directors, responsible for choosing good breeding stock, recommending technical procedures, etc., all of which took some knowledge of science. Where did he learn the knowledge that supported these practices? Partly from experience and concern. However, he also consulted with our local extension agent, and although I seldom saw him reading (he was usually too busy). In later years, after I moved to Ohio, I learned that he knew about Louis Bromfield. He had read his books, especially those that discussed Bromfield's theories of conservation farming developed on his farm not far from my home in Columbus.

If my father, with a minimum of formal education, could become an applied scientist and conservationist, why not our politicians, lawyers, business people, common everyday citizens? Those schooled in what is generally considered the best university system in the world? I suspect with the proper education in science, such as effective global science literacy programs, they could become well informed in science and of the knowledge science develops concerning our habitat. Education has become the substitute for the kinds of practical experiences formerly shared by my father and his sons. Thus, science education has a fundamental role in providing the experiences and knowledge that will lead to an effective understanding of the Earth system we all share, and that our descendants will inherit. However, it must be the right type of science education. We hope that this book and its focus on Global Science Literacy can be a contribution toward providing the "right" science education.

Acknowledgements

This project started with a discussion on Global Science Literacy with Barbara Klemm in a hotel lounge on the beach at Waikiki, Hawaii. She suggested that we put together a book that would spell out the basis and philosophy of the concept. I am thankful for her idea and the constant interest and encouragement she has provided during this project. I also appreciate the enthusiasm expressed by each of the authors and the quality of their composing and writing efforts. Almost all of our communications were accomplished over the Internet via a GSL homepage and email. The authors' responses to editing requests, and suggestions for modifications were always prompt and accurate.

A modest amount of funding was available for the actual production of the manuscript and a presentation on the book at a conference of the National Association for Research in Science Teaching. The funds came from the Alphyl Memorial Fund at The Ohio State University. This is a fund established by students and colleagues to support activities in Earth Systems Education. It is named in honor of Victor and Phyllis Mayer, my parents. Alphyl is a combination of the first letters of my mother's and her mother's first names. It is the name of the farm in Wisconsin worked by my father and the birthplace and early home of his three sons.

VJM
Columbus, Ohio, USA
November, 2001

Contributing Authors

Rosanne W. Fortner
The Ohio State University,
USA

Masakazu Goto
National Center for
Educational Policy Research,
Japan

Chris King
Keele University, UK

E. Barbara Klemm
The University of Hawaii-
Manoa, USA

Yoshisuke Kumano
Shizuoka University, Japan

Fernando Lillo
Santiago de Compostela
University, Spain

José Lillo
Vigo University, Spain

Jeonghee Nam
Pusan National University,
Korea

Nir Orion
Weizmann Institute of
Science, Israel

Hiroshi Shimono
National Center for
Educational Policy Research,
Japan

William Slattery
Wright State University, USA

David B Thompson,
Keele University, UK

Akira Tokuyama
Hyogo University of Teacher
Education, Japan

Roger David Trend
Exeter University, UK

SECTION ONE: FOUNDATIONS FOR GLOBAL SCIENCE LITERACY

The three chapters in this initial section of the book, establish the philosophical, scientific, and historical background for "Global Science Literacy (GSL)." GSL was developed in a period during which a great deal of time, money and effort was devoted to the renewal of science curricula in the United States and worldwide. Conducted in an era dominated by the Cold War and economic competition between nations, very little of that effort addressed the current and likely future directions of science following the end of the Cold War and the onset of Globalization. We believe it is time to rethink the aims and objectives of science curricula especially those at the secondary school level with an eye to the future instead of the relying on current and past applications of science as the basis of science restructuring activities. This is true, we believe, not only in the United States and Japan, but worldwide.

Chapter One reviews the development and rationale of the curriculum effort of Earth Systems Education (ESE). Stimulated by the science report *Earth System Science*, published in 1986, ESE became a curriculum effort in which secondary school science curricula were organized conceptually rather then by the traditional disciplines. The concept is that of the Earth system, in reality, the subject of study of all sciences. The chapter then continues by describing the combination of ESE with objectives for the social studies construct of global education resulting in a form of science curriculum thought useful to all nations, certainly those developed countries with a democratic system of governance. We conclude by applying this construct to the educational system of the nation of Japan.

In Chapter Two we document the historical relationship between national priorities for science and their link to the nature of a nation's science curriculum. We document changing national priorities for science in the United States over the past century and one-half. We then draw a correlation between those changing priorities for science and the science curriculum. The current science curriculum focuses its design and product on the physical sciences. It downplays, if not ignores, the system sciences that will be crucial in the solution of those environmental and social problems left as a residue of the Cold War and intense economic competition.

In Chapter Three we describe the continuum of science methodology from reduction science on one end to that of the system sciences on the other and we argue for the inclusion of the methodology of the system sciences in secondary science curricula. The chapter describes in some detail the nature of system science methodology and cites examples of that methodology and its success in describing complex aspects of nature. We also suggest that system science methodology is culturally tuned to Eastern thinking, in a way that reduction science is not. Therefore it can provide the basis for a science curriculum that more closely reflects the cultural values of Eastern countries.

The science curriculum can be an effective vehicle for increasing communication and understanding within and among societies throughout the world. Such curricula can provide our citizens with a practical understanding of the nature and value of modern science and its ability to help solve many of the environmental

1

V.J. Mayer (ed.), Global Science Literacy, 1–2.
© 2002 *Kluwer Academic Publishers. Printed in the Netherlands.*

and social problems now confronting the stability and livability of the world's nations. To be this vehicle, however, secondary school science students around the world, must be very familiar with the nature and operation of the system that they are a part of, the Earth system, and the methods that scientists have used todiscover its history and its processes. The three chapters in this section of the book, lay the groundwork for a curriculum effort that we believe will be effective, not only in accomplishing those goals, but in eliciting a greater interest in science and its product among secondary school students of science.

CHAPTER 1: EVOLUTION OF GLOBAL SCIENCE LITERACY AS A CURRICULUM CONSTRUCT

Victor J. Mayer
The Ohio State University, USA
and
Akira Tokuyama
Hyogo University of Teacher Education , JAPAN

1. INTRODUCTION

Science as a major component of school curricula can provide a model of a process for effective communication and decision making across barriers to understanding imposed by differences in culture and language, a fundamental objective of the social studies curriculum construct of global education (Anderson, 1992). The scientific process can provide a model for achieving dialogue among peoples with different languages and from diverse cultures. Science in school curricula can also become a common meeting ground for science teachers and social studies teachers. It provides an avenue of linkage between the curricular areas and an opportunity for interdisciplinary planning and teaching. Together social studies and science teachers can help to ensure that our future leaders and voters will understand our interrelationships with peoples around the world and how our daily activities affect our planet and its resources. This is a fundamental goal of global education. It also lies at the core of Earth Systems Education, discussed in the next section of this chapter, and our efforts to develop a global science literacy rationale and program, the topic of this chapter.

Scientists throughout the world have a single shared subject—Earth and its environment in space--and a common method of study and communication--the procedures and language of science. Scientists start with an accurate description of their observations about Earth processes and materials and then go on to develop logical arguments and interpretations based on those observations of nature. But also, science is a collective endeavor. Examine, for example, the authorship credits for a recent research article in one of the premier American scientific journals, *Science*, published by the American Association for the Advancement of Science. It is likely that the article will have several authors from different countries or at least from different cultural heritages. In addition, it will cite the work of many others in the same or related fields of science again often from different countries or cultural heritages. These individuals use the mental processes of science and its language to communicate across their disparate cultural experiences and identities in solving problems or studying processes occurring in our Earth systems.

3

V.J. Mayer (ed.), Global Science Literacy, 3–24.
© 2002 *Kluwer Academic Publishers. Printed in the Netherlands.*

Often a science report will challenge the previous work of those cited. Such a challenge is what keeps science honest even though some of its participants may not be. A scientist's work is replicated, or at least reevaluated, by others and is therefore subject to change and reinterpretation or even outright rejection by the scientific community. As individuals, scientists possess all of the frailties of humans. They make mistakes. They might even misrepresent their data. But if so--they will be found out and corrective action taken. The result of these procedures of science, therefore, is a representation of an Earth process or material that has a high probability of representing that aspect of the real world. Scientists throughout the world, always check their work against the same standard, the observations they make of Earth and its environment in space. Thus, science and its subject--the Earth system--provide an international avenue for communication across the barriers imposed by language and culture. As such the methods of science provide a model of a process for the honest evaluation of social issues and decisions on governmental policies and potential actions that can be less susceptible to bias than are other forms of decision making. It provides a model for an individual's evaluation of information received from a variety of sources and a mechanism for making informed decisions in this ever more complex world.

2.   DEVELOPMENT OF EARTH SYSTEMS EDUCATION AS THE SCIENCE FOUNDATION OF GSL

In the past twenty years there have been tremendous advances in the understanding of planet Earth from the application of advanced technology in data gathering by satellites and data processing by supercomputers. As a result, Earth scientists have reinterpreted the relationships between the various subdisciplines and their mode of inquiry. These changes are documented in the "Bretherton Report," developed by a committee of scientists representing various American government agencies with Earth science research mandates. The committee was chaired by Francis Bretherton, a meteorologist at the University of Wisconsin (Earth System Sciences Committee, 1988). This reconceptualization of the process and goals for study of planet Earth was called 'Earth System Science'.

We believe that this report establishes a basis from within the science establishment for conceptually organizing science curricula. Earth system science has now become a model for much of the geoscience research carried on in the USA, not only by government agencies, but by academic institutions and industry as well. Earth system science, instead of the discipline oriented approach to the study of the atmosphere, biosphere, hydrosphere, lithosphere, the solar system and the universe, takes an interdisciplinary or conceptual approach. Physicists, chemists, biologists, geologists--scientists from many different disciplines including the social sciences--work cooperatively applying their special knowledge, skills, and methodology to understand how each of the Earth systems work, how they interact, and how humans affect those systems.

Earth processes (taken to include those operating within the Earth system, the solar system and the universe at large) to be the subject of current and future research were divided by the committee into two time scales. One deals with relatively short term processes such as those of weather, climate change, and nutrient cycles. These are the processes potentially influenced by human behavior and occur in relatively short time frames from seconds to hundreds of years. The other time scale includes long term processes, such as plate tectonics and the evolution of life operating over thousands to millions of years. The short-term processes are of special concern to the world community because of the disturbances introduced into the Earth systems over the past century by the invention and application of many technologies and by the rapidly growing world population. An understanding of the long-term processes provides a philosophical place for the human presence within the Earth system. Such a background would assist students to more easily comprehend those essential contributions of Copernicus, Galileo, Hutton, Darwin and others that describe our place in the universe.

The committee defined Earth system science with seven statements (ESSC, 1988, p. 21). The first two contrast the traditional Earth science view with that of the new Earth systems science view.

- The two traditional motivations for Earth science are an understanding of the Earth as a planet and the search for practical benefits from such research.
- Earth system science treats the Earth as an integrated system of interacting components, whose study must transcend disciplinary boundaries.

The third points out the reason for this changing focus of the sciences.

- Earth system science has been stimulated by the maturation of the traditional disciplines, a global view of the Earth from space, and the increasing role of human activity in global change.

The next two points explain the two divisions of processes the committee defined.

- On time scales of thousands to millions of years, Earth processes are driven both by internal energy and the external energy of solar radiation.
- On time scales of decades to centuries, Earth processes are dominated by the physical climate system and the biogeochemical cycles, with human activities playing an increasing role in both.

The goal of Earth system science is:

- .... to obtain a scientific understanding of the entire Earth system on a global scale by describing how its component parts and their interactions have evolved, how they function, and how they may be expected to continue to evolve on all time scales.

The committee sees the challenge of research to be to:

- .... develop the capability to predict those changes that will occur in the next decade to century, both naturally and in response to human activity.

Except for the last point, these statements could be taken to define the nature of a secondary school science curriculum, not just the Earth science curriculum often offered in the ninth grade of American schools.

A second report, dealing only with the solid Earth sciences makes the following recommendation:

> **Efforts need to be made to expand Earth science education to all.** Citizens need to understand the Earth system to make responsible decisions about use of its resources, avoidance of natural hazards, and maintenance of the Earth as a habitat. School systems must respond to this need.... (National Research Council, 1993, p. 12)

This report links the need for better understanding among our citizens to the important needs of society in this post-cold war era. As such it provides important support for readdressing the importance of including significant content about the Earth system in the nations's science curricula.

3. EARTH SYSTEMS EDUCATION

Earth system science can provide science educators with a conceptual approach to curriculum integration as suggested by Mayer (1995) in proposing the curriculum design effort called Earth Systems Education, not just for a narrowly defined Earth science course, but for the entire secondary school science curriculum. We suggest that such a curriculum could replace the "layer cake" of Earth science, biology, chemistry and physics in the United States and those separate courses taught at various grade levels in other countries. Using the concept of the Earth system and its processes as the organizing framework, basic physical and chemical principles can be learned by the student in a context of intimate importance, the student's habitat. Thus the basic principles of science can be taught more meaningfully and thus be more easily understood and retained. Using such a conceptual approach to the organization of curricula might also avoid one of the fatal elements in past attempts to integrate the science curriculum, the competition between representatives of each of the science disciplines for their 'rightful' place within the curriculum.

When it appeared that science curriculum restructuring efforts in the United States might once again ignore planet Earth, a conference of geoscientists and educators was organized by the American Geological Institute and the National Science Teachers Association with support from the National Science Foundation. It took place in Washington, DC in April 1988. The forty scientists and educators, including many scientists from the agencies responsible for the Bretherton Report, met over a period of five days. Through small group interaction techniques they developed a preliminary framework of four goals and ten concepts from the Earth sciences that they felt every citizen should understand (Mayer and Armstrong, 1990). Through the work of the conference participants and subsequent discussions with teachers and Earth science educators at regional and national meetings of the National Science Teachers Association, a new focus and philosophy for science curriculum emerged under the label, Earth Systems Education (Mayer, 1991).

In Spring of 1990, the Teacher Enhancement Program of the National Science Foundation awarded a grant to The Ohio State University and the University of Northern Colorado for the preparation of leadership teams in Earth Systems Education--PLESE, the Program for Leadership in Earth Systems Education. The objective of the program was to infuse more content regarding the modern understanding of planet Earth into the nation's K-12 science curricula. In preparation for this program, the PLESE planning committee met in Columbus in May 1990, to develop a conceptual framework which would be used to guide the content and philosophy of the program. Input for their work included the Project 2061 report (AAAS, 1989) and the results of the April 1988, conference. Over a period of five days the committee developed a Framework for Earth Systems Education consisting of seven understandings (see figure 1). These understandings provided a basis for the PLESE teams to construct curriculum guides for their areas of the country and for selection of existing materials for implementing Earth systems education in their areas. The PLESE Planning Committee intentionally arranged the understandings into a sequence to draw attention to the importance of the first two understandings especially since they are seldom if ever given importance in traditional science curricula.

Table 1. Framework For Earth Systems Education

Understanding #1: Earth is unique, a planet of rare beauty and great value.

- The beauty and value of Earth are expressed by and for people through literature and the arts.
- Human's appreciation of planet Earth is enhanced by a better understanding of its subsystems.
- Humans manifest their appreciation through their responsible behavior and stewardship of subsystems.

Understanding #2: Human activities, collective and individual, conscious and inadvertent, affect planet Earth.

- Earth is vulnerable, and its resources are limited and susceptible to overuse or misuse.
- Continued population growth accelerates the depletion of natural resources and destruction of the environment, including other species.
- When considering the use of natural resources, humans first need to rethink their lifestyles, then reduce consumption, then reuse and recycle.
- By-products of industrialization pollute the air, land, and water, and the effects may be global as well as near the source.
- The better we understand Earth, the better we can manage our resources and reduce our impact on the environment worldwide.

Understanding #3: The development of scientific thinking and technology increases our ability to understand and utilize Earth and space.

- Biologists, chemist, and physicists, as well as scientists from the Earth and space science disciplines, use a variety of methods in their study of Earth systems.
- Direct observation, simple tools, and modern technology are used to create, test and modify models and theories that represent, explain, and predict changes in the Earth system.
- Historical, descriptive, and empirical studies are important methods of learning about Earth

and space.
- Scientific study may lead to technological advances. Regardless of sophistication, technology cannot be expected to solve all of our problems.
- The use of technology may have benefits as well as unintended side effects.

Understanding #4: The Earth system is composed of interacting subsystems of water, rock, ice, air, and life.

- The subsystems are continuously changing through natural processes and cycles.
- Forces, motions and energy transformations drive the interactions within and between the subsystems.
- The Sun is the major external source of energy that drives most system and subsystem interactions at or near the Earth's surface.
- Each component of the Earth system has characteristic properties, structure, and composition that may be changed by interactions of subsystems.
- Plate tectonics is a theory that explains how internal forces and energy cause continual changes within Earth and on its surface.
- Weathering, erosion, and deposition continuously reshape the surface of the Earth.
- The presence of life affects the characteristics of other systems.

Understanding #5: Planet Earth is more than 4 billion years old and its subsystems are continually evolving.

- Earth's cycles and natural processes take place over time intervals ranging from fractions of seconds to billions of years.
- Materials making up planet Earth have been recycled many times.
- Fossils provide the evidence that life has evolved interactively with Earth through geologic time.
- Evolution is a theory that explains how life has changed through time.

Understanding #6: Earth is a small subsystem of a solar system within the vast and ancient universe.

- All material in the universe, including living organisms, appears to be composed of the same elements and to behave according to the same physical principles.
- All bodies in space, including Earth, are influenced by forces acting throughout the Solar System and the universe.
- Nine planets, including Earth, revolve around the sun in nearly circular orbits.
- Earth is a small planet, third from the Sun in the only system of planets definitely known to exist.
- The position and motions of Earth with respect to the Sun and Moon determine seasons, climates, and tidal changes.
- The rotation of Earth on its axis determines day and night.
-

Understanding #7: There are many people with careers that involve study of Earth's origin, processes, and evolution.

- Teachers, scientists, and technicians who study Earth are employed by businesses, industries, government agencies, public and private institutions, and as independent contractors.
- Careers in the sciences that study Earth may include sample and data collection in the field and analyses and experiments in the laboratory.
- Scientists from many cultures throughout the world cooperate and collaborate using oral, written, and electronic means of communication.
- Some scientists and technicians who study Earth use their specialized understanding to locate resources or predict changes in Earth systems.
- Many people pursue avocations related to planet Earth processes and materials.

The first understanding emphasizes the aesthetic values of planet Earth as interpreted in art, music and literature (Mayer, 1989). It stresses the creativity of the human spirit and how that creativity has perceived and represented the planet on which we live; a creativity that is also essential to the proper conduct of science. This understanding is based on the first concept concerning science education developed by the scientists at the 1988 conference. They felt that the aesthetics of the Earth system belonged in science programs, since it was often the appreciation and awe of the beauty inherent in an Earth system that caused them to become scientists. An appreciation of the Earth system naturally leads the student into a concern for the proper stewardship of its resources; the second understanding of the framework. Orion, in Chapter Eleven, describes the 'green paradigm'as one of three that are heavily influencing the science curricula of several Western countries. A developing concern for conserving the economic and aesthetic resources of our planet leads naturally into a desire to understand how the various subsystems function and how we study those subsystems; the substance of the next four understandings. In learning how the subsystems function students must master basic physics, chemistry and biology concepts. The last understanding deals with careers and avocations in science bringing the focus once again back to the immediate concerns and interests of the student.

The PLESE program extended over a period of five years through the cooperation of faculty at the Ohio State University and the University of Northern Colorado (Mayer, Fortner and Hoyt, 1995). Over 200 teachers and administrators were involved in the program as participants and contributors to the formulation of the concept of Earth Systems Education. A number of Earth Systems Education school programs around the country were developed in direct response to the program. A steering committee was identified from program participants. It met in Spring of 1995, and produced a report guiding the future direction of ESE as a foundation for integrated curriculum development. The ESE steering committee, comprised of teachers and educators representing all of the various sciences, recommended the Earth Systems Education Framework as a foundation for the development of integrated science curricula.

The ESSC has provided a start for such curriculum restructure. The committee devised two 'concept maps.' One deals with the short-term processes and the second with the long-term processes (ESSC, 1988, pp. 27-29). We would suggest, for example, that a simplified version of the Earth system science concept map of short term processes could provide a basis for structuring a two-year curriculum encompassing grades eight and nine, whereas a simplified version of the long term process concept map could provide the curricular basis for grades ten and eleven. Grade twelve science could then be devoted to a study of specific science disciplines allowing those students interested in pursuing science as a career to obtain an understanding of the science discipline and its methodology that interests them. The short term process map can be found at the following Internet address: www.usra.edu/esse/BrethColor.GIF. Unfortunately the maps are in color and much too large to be included with this chapter.

Several development projects have been completed for middle school and high school curricula and in teacher education that reflect an Earth systems approach. One project, developed by teachers in two high schools in a suburban community, substituted a two-year long integrated 'Biological and Earth Systems Science' course for the traditional 9th grade Earth science and 10th grade biology courses (Fortner, et al, 1992). As a result of the Program for Leadership in Earth Systems Education (PLESE) (Mayer, Fortner, Hoyt, 1995) funded by the National Science Foundation, a resource book, including the contributions from nearly 200 teachers involved in the PLESE program over a five-year time period, is available (Mayer and Fortner, 1995). It has been used in several National Aeronautics and Space Administration education programs and several undergraduate teacher education programs. ESE is also seen as a science curriculum entry point for environmental education (Fortner, 1991; Orion, 1998) because of the stewardship components that emerge from thoughtful consideration of human effects on Earth systems. Mayer has contributed a chapter to a handbook for teacher educators on the teaching of global perspectives. His contribution focuses on the use of the science curriculum in reaching certain objectives of global education (Mayer 1997a). Fortner, working with numerous formal and non formal educators, has developed teacher education materials for Earth Systems Education about current conditions in the Great Lakes region. Curriculum units on the wreck of the Edmund Fitzgerald, the origins and impacts of non indigenous species, the changing shape of Niagara Falls, and the impact of water levels on wetland organisms are among ten books of secondary science curriculum activities described at the ESE web site (http://earthsys.ag. ohio-state.edu).

We trust that these projects and similar ones to be conducted in the future, will make a major contribution to refocusing our science curriculum on current and future priorities of science and will indeed help to influence those priorities not only in the United States but in democratic countries worldwide.

4. THE NATIONAL SCIENCE EDUCATION STANDARDS AND GLOBAL SCIENCE LITERACY

We believe that the most significant and far reaching effort to define the essentials of science literacy in the United States was that of the National Academy of Sciences in developing the National Science Education Standards (NRC, 1996). After three years of deliberation by three committees comprised of science educators, teachers and scientists, and the submission of three successive drafts of the report to over 10,000 American teachers and scientists for review and comment, the National Research Council of the Academy published its report in 1996. It represents a very hard to reach consensus of the various stakeholders in science education in the United States and therefore has a significance beyond previous efforts in attempting to define the nature of science for school curricula. Although there are six areas of standards we believe that the first four have particular relevance for developing international standards of science literacy. They are also the ones that we have directly incorporated into Earth Systems Education which in turn forms the foundation for Global Science Literacy.

1. **Science Teaching**. These standards and their placement at the beginning of the report are a recognition that teachers are the essential element in accomplishing effective science education. It is they who are responsible for establishing a climate of inquiry in the classroom, the essential theme which runs throughout the standards. They emphasize that "inquiry into authentic questions generated from student experiences is the central strategy for teaching science" (p. 31). We have incorporated science teaching methods into Global Science Literacy that are consistent with establishing the climate of inquiry emphasized in this standard. They include:

- Cooperative learning: We suggest that teachers need to change their reliance on lecture, discussion, textbooks and structured laboratories. Students must see themselves as fundamentally responsible for their learning. Perhaps the most effective procedure that we have found for this purpose is cooperative learning, especially that form called the 'jigsaw'. In Chapter Five, Fortner discusses her uses of this technique in her college classes and then applies her experiences in discussing the potential and difficulties of using cooperative learning in the secondary school classroom.

- Fieldwork: Extensive fieldwork in the local community should be the basis of much of science learning. It provides a primary source of data for use in investigating the Earth system. In Chapter Eight, Shimono and Goto report on a study conducted by a student in Japan assisted by her parents and in Chapter Fourteen, Goto describes his work in incorporating fieldwork into the science curricula of a lower secondary school also in Japan. Thompson in Chapter Eight provides a model of a narrative of the life of a famous Earth system scientist, Darwin, and associated field experiences in the region in which the scientist lived.

- The Internet as a resource: In order to conduct investigations of regional or global earth processes, students can use the Internet as a source of original data in a form that is useful to them. Slattery, Mayer and Klemm in Chapter Six discuss this and other uses of the Internet in Global Science Literacy curricula.

- Reading activities: Not all important science concepts are subject to laboratory or field investigation. Explanatory stories, as proposed by the British educators responsible for compiling the report *Beyond 2000: Science Education for the Future*, (Millar and Osborne, 1998) can be used by teachers to explore the types of inquiry conducted by Earth system scientists in regions such as the arctic, oceans, mountain ranges and jungles that are normally inaccessible to students. King, in Chapter Four, provides examples of an explanatory story and how teachers might use it in helping students understand difficult science concepts. Lillo and Lillo in Chapter Nine discuss the use of ancient writings concerning environmental changes and how they can be developed into useful reading activities.

2. **Professional Development of Teachers of Science**. Following the Teaching Standards is a series of guidelines to assist teacher education programs for preparing science teachers to deliver this enhanced version of science education. They emphasize that "prospective and practicing teachers must take science courses in which they learn science through inquiry...." (NRC, 1996, p. 60). In a future volume, we plan to have authors discuss their experiences with developing Earth systems courses, often taken by teachers at the university level. We will also propose ideas, problems and strategies for developing teacher education programs for Global Science Literacy courses. Teacher education is the least developed of the areas of concern in implementing GSL curricula in schools. It is also the most difficult. Aside from a few universities developing Earth system course sequences, little has been done and knowing the political and economic structure of universities, little can be done in the immediate future to design and implement teacher education programs that would support the introduction of GSL based courses in secondary schools. Instead major efforts need to be directed toward in-service education using models such as that reported in Mayer, Fortner and Hoyt (1995) and those used by the Center for Educational Technology of Wheeling University as reported on in Chapter Six.

3. **Assessment in Science Education**. It is widely recognized that the nature of assessment or testing is one of the most powerful determinants of classroom climate and practice and of the content of the curriculum. This is dramatically illustrated by the results of a study of GSL curriculum implementation in Japan reported in Chapter Fifteen. For change to occur, therefore, the NRC committee responsible for assessment recognized that there must be dramatic change in the nature of assessment practices in both the schools and in national testing programs. They emphasize that "the choice of assessment form should be consistent with what one wants to measure and infer" (p. 85). They also recommend that "eliciting and analyzing explanations are useful ways of assessing science achievement" (p. 92). Nam and Mayer in Chapter Ten discuss this type of authentic and performance assessment, applied to GSL curricula, in the context of Korean science education.

4. **Science Content Standards**. These are the most detailed of the standards, and they were the most controversial and most difficult to achieve consensus. They imply a considerable change in the nature of content in the science curriculum, especially in the courses taught at the high school level in the United States. They advise a considerable shift away from the current narrow focus on the traditional disciplines of biology, chemistry and physics and the current extensive treatment of large volumes of detailed information from those disciplines that characterize our pre-college science courses. Instead, the standards emphasize a more adequate representation of the nature of the scientific enterprise, of the breadth of knowledge developed by science, and of the full spectrum of scientific activity and knowledge including content from the Earth sciences. Table 2 includes a list of the areas of the content standards (p 104-112).

Table 2: National Science Education Content Standards

1. Unifying concepts and processes in science
   a. Systems, order, and organization
   b. Evidence, models, and explanation
   c. Change, constancy, and measurement
   d. Evolution and equilibrium
   e. Form and function
2. Science as inquiry

3. Physical science

4. Life science

5. Earth and space science

6. Science and technology

7. Science in personal and social perspectives

8. History and nature of science.

Later in this chapter, we provide a detailed examination of these standards and how they can be formatted in a Global Science Literacy context.

5 and 6. **Science Education Program and System Standards**. The last two sets of standards are grouped together for brevity and convenience. They both concern the educational systems that exist outside of the science classroom at the local school system level, the state level and the national level. For classroom science to change significantly, each of these levels of administration in turn must change their educational policies. They state that "for schools to meet the Standards, student learning must be viewed as the primary purpose of schooling, and policies must support that purpose" (p. 232). Although these are very important standards to be considered in any fundamental change of science curriculum, it is beyond the scope of this book to discuss them as they might be applied in different countries in the context of Global Science Literacy.

5. GLOBAL EDUCATION

Global education has a long history in American social studies education. Although it started in the 1930s as an interest in assisting students in understanding the history and governance of other countries, it received heightened emphasis and interest following World War II and during the ensuing Cold War. It is an approach to curriculum development that includes elements fostering within students a global perspective including an understanding and appreciation of other cultures. This resurgent movement in social studies following the war stimulated a great deal of controversy especially from among the more conservative members of the American establishment. A very successful National Science Foundation supported curriculum project, for example, developed curricula for upper elementary schools that focused on providing an understanding and appreciation of the cultures and societies of other nations and how they developed. Called *Man a Course of Study* it became a major factor in the termination of NSF curriculum projects at the beginning of the Reagan Administration. Critics charged that the curriculum taught humanism and cultural relativism. With the globalization of the economy, the end of the Cold War and other events of the 1990s, however, there has been a resurgence of interest in and the need for global education programs (Smith, 1994).

The 1991 Yearbook of the Association for Supervision and Curriculum Development entitled *Global Education: From Thought to Action* provides an overview of global education. It is the result of a conference held in 1989, organized by the Center for Human Interdependence at Chapman College. The participants at the conference developed the following definition of global education, which was then used in writing a series of essays, reflecting the views of conference participants that comprise the chapters of the yearbook:

> Global education involves learning about those problems and issues that cut across national boundaries, and about the interconnectedness of systems--ecological, cultural, economic, political, and technological. Global education involves perspective taking-- seeing things through the eyes and minds of others--and it means the realization that while individuals and groups may view life differently, they also have common needs and wants. (Tye, 1991, p. 5)

The conference staff further elucidated the nature of global education by citing five interdisciplinary dimensions stated by Robert Haney in his 1976 book, *An Attainable Global Perspective*.

- *Perspective consciousness*: An awareness of and appreciation for other images of the world.
- *State of the planet awareness*: An in-depth understanding of global issues and events.
- *Cross-cultural awareness*: A general understanding of the defining characteristics of world cultures, with an emphasis on understanding similarities and differences.
- *Systematic awareness*: A familiarity with the nature of systems and an introduction to the complex international system in which state and non-state actors are linked in patterns of interdependence and dependence in a variety of issue areas.
- *Options for participation*: A review of strategies for participating in issue areas in local, national, and international settings. (Lamey, p. 53)

Anderson also participated in this conference and contributed a chapter on *A rationale for global education*. In the introduction to a later volume (1992), he summarized four goals of global education. It is a curriculum that engages students of all ages and in all subject matters in the study of:

> ...humankind as a singular entity interconnected across space and time.
> ...the earth as humankind's ecological and cosmic home.
> ...the global social structure as one level of human social organization.
> ...themselves as members of the human species, as inhabitants of planet earth, and as participants in the global social order. (p. xvii)

Some goals of global education, those dealing with planet Earth and its systems, are already implicit in ESE. Accomplishing the remaining, culturally oriented goals can be facilitated by an understanding and use of a modern adaptation of the methods of science as mechanisms for greater communication and understanding between different cultures. In 1996, Mayer accepted a position as researcher in a program in global education supported by the Ministry of Education, Science and Culture of

Japan. The program was based at the Hyogo University of Teacher Education. He and his colleague at the university, Tokuyama, integrated the objectives of global education with those of Earth Systems Education to develop a form of Global Science Literacy (Mayer, 1997b).

6. GLOBAL SCIENCE LITERACY CURRICULUM GUIDELINES

To provide an understandable, coherent, and we hope, useful set of guidelines for developing GSL curricula we have used the content standards of the NRC (p. 103-204) as goals, not because they are American in origin, but because they are succinct yet comprehensive. They also best represent the opinions of the American science and education communities. Each is quite broad, no doubt because of the compromises that were necessary to achieve agreement from a variety of committee members representing individuals from all levels of education and from a wide variety of science disciplines. As such, they should find substantial acceptance among science educators worldwide. We have taken these goals, combined them with the four goals of Global Education proposed by Anderson (1992) and constructed a series of graphic representations of the content of Global Science Literacy curricula. They are displayed in Figures 1 through 4.

In each of the diagrams, we have arranged the standards and goals such that those on the outside of the diagrams and in bold type, are the 'end' goals of the curriculum—the understandings that we hope all science instruction would lead to and finally develop. It is those understandings that all individuals must have to truly understand and appreciate their habitat—the Earth system—and how they interact with that habitat. Those in the middle and in ordinary type are subsumed by the 'end' goals and form a basis for the achievement of those 'end' goals. This gives an order and structure to the two- to three-year-long curricula developed to represent the goals and objectives of Global Science Literacy. Whether they will be useful to curriculum designers is not certain. We do hope that they help others to understand the conceptual nature of Global Science Literacy.

According to the NRC, there are a number of standards that should apply to all of the grade levels and others just for the middle school and high school levels. Figure 1 includes the goals that apply to all grade levels of the science curriculum. The key provided for Figure 1 will also help to understand Figures 3 and 4. As in Figures 3 and 4, we have keyed each goal to the appropriate set of content standards and to the applicable Earth Systems Understanding (Table 1). There also may arise some confusion in terminology. Although we may be playing a bit 'loose' with the meanings of standard and goal, when placing the standards into the Earth systems context we refer to them as goals. Occasionally the terms 'concepts' and 'topics' are used. These refer to more specific understandings that would be components of a particular goal (standard).

	U. Systems, order, and organization ESU 4		
U. Change, constancy, and measurement ESU 3, 4	U. Form and function ESU 4	**G. Students as members of the human species, as inhabitants of planet Earth, and as participants in the global social order ESU 4**	
	U. Evolution and equilibrium ESU 5		
	I. Abilities necessary to do scientific inquiry ESU 3		
	T. Understanding about science and technology ESU 3		
	H. Science as a human endeavor ESU 3, 7		
I. Understandings about scientific inquiry ESU 3			

Figure 1. Possible organization of Unifying Concepts (U), Standards (I = Inquiry, H= History and Nature of Science, T= Science and Technology) and Global Education Goal (G) common to all grade levels for use in developing science curricula. The code 'ESU' refers to the Earth Systems Understanding(s) most related to the particular standard or goal.

The knowledge of our Earth system, experienced by every citizen of Earth, is the central, organizing principle in the remainder of the figures. Thus, each of the content standards developed by the NRC were evaluated for their consistency with Earth Systems Education Understandings (Table 1) and ESE philosophy and arranged accordingly in the diagrams. In Figure 2, we have provided the generalized scheme for the organization of the remaining figures. The Earth and Space Science Standards (E) and the Science in Personal and Social Perspectives (S) were taken as the 'end' goals and others, then, related to them. Earth systems standards are placed at the top and left side. The Personal and Social Perspectives Standards are positioned on the right side. To these we have added goals in global education (G) appropriate for global science literacy at the bottom of each diagram. All of those considered as 'end' goals for Global Science Literacy curricula are highlighted in bold.

	E. Earth System Standards	
E. Earth System	B. and P. Biological and Physical Science Standards	**S. Science Personal Social Perspec**

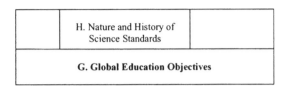

	H. Nature and History of Science Standards	
	G. Global Education Objectives	

Figure 2. General organization of Figures 3 and 4. The letters at the beginning of each statement are used to indicate the group in which the standards and objectives represented in figures 3 and 4 are included. See text for explanation.

E. Structure of the Earth system ESU 4					
E. Earth in the Solar system / **E. Earth in the Solar system ESU 6**	B. Structure and function of living systems ESU 4	P. Transfer of energy ESU 4	P. Properties and changes of properties in matter ESU 4, 5		**S. Natural Hazards ESU 4**
	P. Motions and forces ESU 4, 6				
	B. Populations and ecosystems ESU 4	B. Populations, resources and environments ESU 2			
E. Earth's history ESU 5	B. Reproduction and heredity ESU 4	B. Diversity and adaptations of organisms ESU 4	B. Regulation and behavior ESU 4	S. Personal health ESU 1, 2, 4	**S. Risks and benefits ESU 2, 7**
	H. History of science ESU 3				
	H. Nature of science ESU 3	S. Science and technology in society ESU 2, 3, 7			
	G. Humankind as an entity interconnected across space and time ESU 4, 5, 6	**G. Earth, humankind's ecological and cosmic home ESU 1, 4**			

Figure 3. Possible organization of science standards and global education objectives for use in developing middle school (years 6 to 9) science curricula. Letters at the beginning of each standard identify the group it belongs to (refer to Figure 1) except for "G". It identifies global education objectives. See text for explanation of the diagram.

In Figure 3 we have organized the standards suggested for middle school, and in Figure 4 those for high school.

An illustration of the use of these diagrams can be seen by applying them to the study by Orion in Chapter Eleven. Using Figure 4, the 'end' goal of his units on the water cycle, carbon cycle and rock cycle would be "Energy in the Earth system". This goal would be attained through the understanding of concepts subsumed under "Geochemical Cycles" which in turn would be assisted by the understanding of appropriate topics selected from "Chemical Reactions", "Motions and Forces", "Structure and Properties of Matter". These understandings in turn would lead to the other side of the diagram, with the broad understanding of topics from "Natural Resources" and "Environmental Quality" as additional 'end' goals. This is a "post-analysis" and as such is imperfect but illustrates the potential use of these diagrams or improved versions of them. Klemm in Chapter Twelve provides examples of how to design curriculum experiences for children with special learning difficulties. She relates the learnings to the ESU's.

Although the NAS provided content standards for the primary years, (kindergarten through 4) we have not developed guidelines for global science literacy for those years of schooling. We are not certain that children in that age group have the necessary mental maturity to cope with the global issues inherent in global science literacy. Perhaps the major reason, however, is that the authors of this chapter have no special background in elementary science education. We hope that someone in the future can take on the task of rationalizing the role of Global Science Literacy for the elementary science level.

The result of this work is a framework that can be used to develop programs in global science literacy at the middle school and high school levels. We believe that science curricula organized in this manner will assist school programs in providing citizens with a broader understanding of cultures and assist them in communicating with cultures not their own. The science they learn will also help future citizens understand the role of science in solving environmental and social problems left in the wake of the Cold War as well as contribute to their aesthetic and intellectual appreciation of their habitat

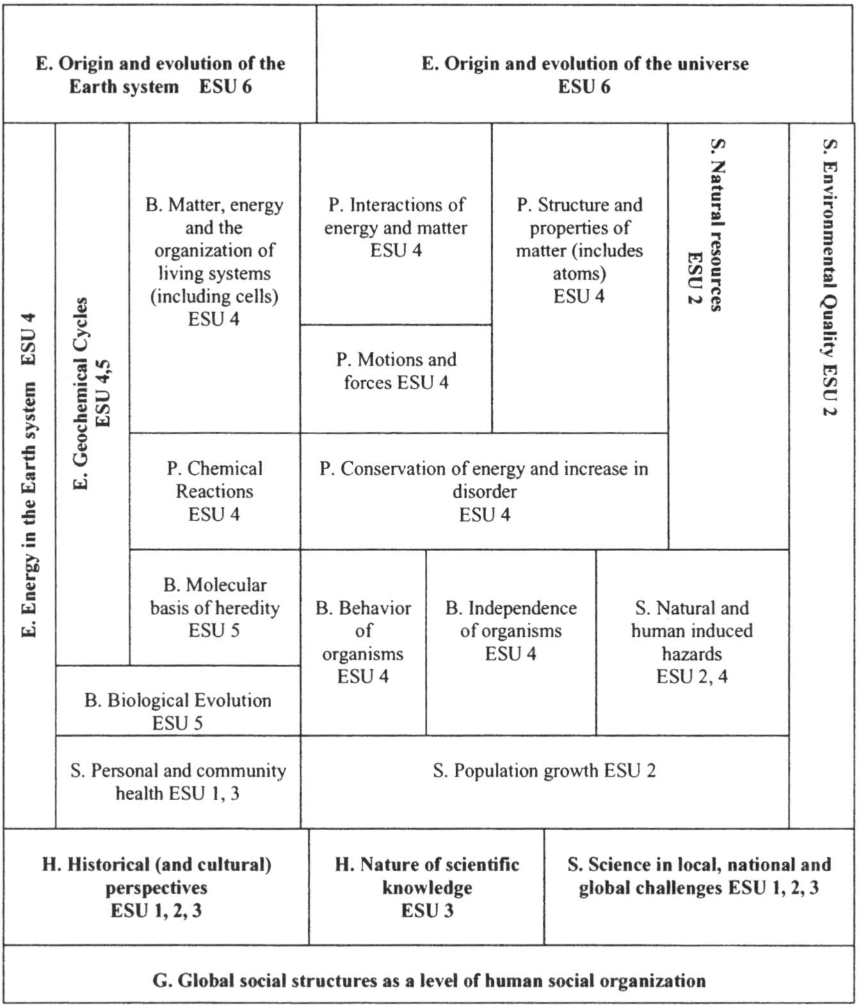

E. Origin and evolution of the Earth system ESU 6	E. Origin and evolution of the universe ESU 6				

Figure 4 diagram content:

E. Energy in the Earth system ESU 4	E. Geochemical Cycles ESU 4,5	B. Matter, energy and the organization of living systems (including cells) ESU 4	P. Interactions of energy and matter ESU 4	P. Structure and properties of matter (includes atoms) ESU 4	S. Natural resources ESU 2	S. Environmental Quality ESU 2
			P. Motions and forces ESU 4			
		P. Chemical Reactions ESU 4	P. Conservation of energy and increase in disorder ESU 4			
		B. Molecular basis of heredity ESU 5	B. Behavior of organisms ESU 4	B. Independence of organisms ESU 4	S. Natural and human induced hazards ESU 2, 4	
		B. Biological Evolution ESU 5				
		S. Personal and community health ESU 1, 3	S. Population growth ESU 2			

H. Historical (and cultural) perspectives ESU 1, 2, 3	H. Nature of scientific knowledge ESU 3	S. Science in local, national and global challenges ESU 1, 2, 3

G. Global social structures as a level of human social organization

Figure 4. A possible organization of science standards and global education objective for high school (years 10 through 12) science curricular development. Letters in front of the topic indicate the group of standards topic is related (refer to Figure 1) to except for the letter 'G'. It indicates a global education objective. The code ESU indicates the Earth Systems Education Understanding the topic is related to. See text for explanation of the diagram.

7. GLOBAL SCIENCE LITERACY AND ITS IMPLICATIONS FOR SCIENCE CURRICULUM IN AN EASTERN COUNTRY: JAPAN

Global Science Literacy, as we have defined it thus far in this chapter, has been based on the American science standards organized according to a conceptual framework consistent with that developed by the Earth System Science Committee and the goals of Global Education. The sources are not only Western, but American. Would the philosophy underlying a curriculum based on these constructs be appropriate for other countries, especially those of Eastern culture and thought? Japan is a country with a culture very different from the predominant Western culture of the USA and the Western culture in which science was developed. According to Kawasaki (1996), science educators in Japan and other non-Western cultures must be aware of the nature of modern science and how it is interpreted in school science programs and the science curriculum implications of the linguistics of non-Western languages.

Modern science evolved out of Western thought and philosophy. It developed in a Judeo-Christian religious system in which nature was believed to have been given to people for their use. The natural environment was often seen as a hostile element to be controlled for the safety and security of people. Thus, a duality in thought about humans and their environment arose that was influential in the early development of science. In this context, humans devised methods to use the resources of the environment to sustain themselves and to protect themselves from nature without concern as to what such use did to the natural systems providing those resources. Science was developed, in part, as an instrument in this struggle with our habitat.

In contrast, Eastern thought, implicit in the history of Japan and in the aesthetic and cultural attitudes of its citizens, holds that people are one with nature. People also rely on nature to sustain themselves, understanding however that humans are integral parts of nature interacting with it to the benefit of both. Nakamura (1980) states that this cultural understanding of unity with nature had its origins in ancient Japan:

> For the ancient Japanese the mythical world and the natural world interpenetrated one another to the extent that human activities were explained and sanctioned in terms of what kami, ancestors or heros, did in primordial time. .... (... the actual national religion of the Japanese has been essentially a nature-religion, with its pantheon consisting of nature-gods.) (p. 243)

This basic philosophy of the oneness of people and nature is reflected in the various festivals held during the year related to natural events, such as the blossoming of cherry trees in spring and the viewing of the moon in September. According to Nakamura,

> This sentiment for nature, which contributed to the sympathetic heart of the Japanese people and their love of order in communal life, may be due partly to the influence of the land and climate and to early attainment of a settled agricultural civilization....

The love of nature, in the case of the Japanese is tied up with their tendencies to cherish miniature forms and treasure delicate things. (p. 249)

Such customs may be held over from an earlier agricultural society, but they also result from the influence of Shintoism and Buddhism--belief systems central to Japanese life and thought. According to Buddhist philosophy, men constitute just one class of living beings. As such, they have no right to unlimited use of natural resources. In addition, men have no right to unlimited exploitation of animal and plant life, which form part of nature. Until recently, Westerners or moderns tended to think that men were quite separate and different from the natural world. This assumption is ungrounded and unreasonable. It has brought about devastation. Now men incur retaliation by nature (p. 300).

Many activities relating to nature are enjoyed by the Japanese (Ekiguchi and McCreery, 1987). The subjects of both historical art and contemporary art often focus on nature with mountains, trees, streams, and again, flowers being favorite subjects. The practice of the ancient art of Japanese garden design continues with local householders, especially in rural areas. The very small compact yards associated with residences are often reproductions of Japanese landscapes, including mounds representing hills and mountains, a small stream or pond representing a river, lake or bay, large rocks and beautifully shaped trees to represent clouds, and a variety of flowers and shrubs, all carefully placed to be in harmony and to produce the total effect of a miniature natural landscape. On a larger scale, publicly maintained gardens in cities such as Himeji, Hiroshima and Kyoto are spectacular in their carefully arranged architecture. Gardens on the grounds of shrines are designed to reproduce in miniature a famous Japanese landscape (Nakamura p. 253). Discussions with Korean science educators in a seminar held during the Tenth Anniversary meeting of the Korean Association for Research in Science Education in 1996, confirmed that this thought system affirming the oneness of nature and people and the intimate and sustaining relationships between them is shared to varying extent by other Asian nationalities, perhaps as a result of the influence of Buddhism.

8. CORRESPONDENCE OF SYSTEM SCIENCE AND GLOBAL SCIENCE LITERACY TO EASTERN THOUGHT

Science as it is taught in most countries including Japan is based on Western practice (Kawasaki, 1996), especially that in the USA. Science educators in non-Western countries must be able to adapt a science based in Western philosophical traditions, and curricula based in Western practice and language to the culture and language of their particular countries. We contend however, that using a Global Science Literacy approach and incorporating the cultural and aesthetic considerations implicit in Japanese culture as 'border crossings' into science suggested by Aikenhead (1996), curriculum specialists can design science curricula closely related to Japanese culture.

Aspects of Eastern thought are supportive of the systems approach in science that has evolved through the Earth sciences and ecology. We develop the reasoning behind this statement in Chapter Three. Curriculum programs based conceptually and methodologically on Earth system science should be found to be much more compatible with the culture and society of Japan and other Asian countries than traditional science programs with their focus on reduction forms of science. Eastern thought, in fact, with its emphasis upon the co-dependence of people and nature and an appreciation of the aesthetic aspects of nature, provides a firm philosophical foundation for curriculum innovations such as Global Science Literacy.

Conceptually, also, Global Science Literacy accommodates an Eastern view of nature. The first ESE Understanding emphasizes the beauty that people find in nature. They suggested this understanding as an important outcome of a science program. The fifth ESE Understanding emphasizes the interactions between the Earth systems. Eastern thought recognizes the interdependence of the human subsystem and the life, water, solid Earth and air subsystems. The recognition of the first and fifth ESE Understandings implies the importance of stewardship of the natural environment, the subject of ESU Two, an importance also supported in Eastern belief. Both ESE Understandings One and Two are thought to be marginal to the practice and product of Cold War science and generally not included in science education programs In fact, the NRC Standards do not recognize the aesthetics found in nature as being appropriate as a component of science curricula.

Those who live in Japan or who have visited there will note that even though Eastern belief systems should support effective environmental policy, this is not the case in practice, especially along the Tokaido corridor. The failure to implement effective environmental policies in heavily developed areas of Asia demonstrates the failure of a science based on the physical science model that is almost solely applied to the development of technology. What is so dramatically true of over development in the East, of course, is also true in the West. What has happened to the environment in countries such as South Korea, China, Taiwan and Japan, however, can be seen as resulting from an uncritical acceptance by the governmental, commercial and industrial sectors of the predominant emphasis in Western science on a science methodology and thought system focused on the development of technology, ignoring the impact of its use on the Earth system.

The Ministry of Education, Science and Culture of Japan (Monbusho) develops educational policies and guidelines for curriculum and teaching. In its report on the conditions of education published in 1994, it makes several recommendations to solve problems it found from a survey of teachers, administrators and parents. First, programs need to be developed that will enhance the goal of internationalization of education. Second, schools need to develop effective environmental education programs. Third, school programs need to be designed to improve the attitudes of students toward science and technology and hopefully, school. We contend that science programs developed following the guidelines of Global Science Literacy can be effective in accomplishing those goals. Mayer, Shimono, Goto and Kumano in Chapter Fifteen report on a research study in which secondary school science teachers in two regions of Japan evaluated the relative effectiveness of GSL in accomplishing the goals stated in the Monbusho report.

9. CONCLUSION

If indeed science is a culture different from that of ethnic cultures, even, as some claim, the Western culture it rose from, then as we suggest here, starting with the student's local natural environment when introducing new science concepts; using it and the global system as a context for all science instruction; and integrating aspects of the student's predominant culture through its art, music and literature, can provide the border crossings into the science sub-culture discussed by Aikenhead (1998). In this manner, both Western and Eastern students might be more likely to see the relevance to their own lives of the processes and information incorporated into the science curriculum. In addition, science as a central component of a school curriculum, designed around Global Science Literacy, can assist in achieving the objectives of the social studies effort of global education. It provides a subject in common with all cultures, the Earth system, and a common method of studying and communicating about that subject.

AKNOWLEDGEMENTS

The foundation for the ideas in this chapter was established during the Program for Leadership in Earth Systems Education, funded by the National Science Foundation from 1990 through 1995 with grants to The Ohio State University and the University of Northern Colorado. It includes contributions from some 200 teachers, administrators and college educators. Recent developments were accomplished in part while the first author was Visiting Research Scholar at the Hyogo University of Teacher Education, Center for Educational Research, JAPAN in 1996, and subsequently in 1998 as Senior Fulbright Researcher attached to Shizuoka University and the Institute for Research in Education of the Ministry of Education, Science and Culture, JAPAN. Additional contributions were made through the Ohio Sea Grant Program.

REFERENCES

Aikenhead, G. S. (1996). Science education: Border crossing into the subculture of science. *Studies in Science Education*, 27, 1-52.
American Association for the Advancement of Science (1989). *Science for all Americans*. Washington, D. C.: AAAS.
Anderson, L. F. (1992). Introduction. In B.B. Tye and K.A. Tye. *Global education: A study of school change*. Albany: State University of New York Press.
Earth System Science Committee (1987). *Earth System Science: A program for global change*. Washington: National Aeronautics and Space Administration.
Ekiguchi, K. and McCreery, R. S. (1987). *A Japanese touch for the seasons*. Tokyo: Kodansa International.
Fortner, R. W. (1991). A place for EE in the restructured science curriculum. In J. H. Baldwin, (ed.), *Confronting environmental challenges in a changing world*. Troy, OH: North American Association for Environmental Education.
Fortner, R. W., Pinnicks, R., Shay, E., Barron, P., Jax, D., Steele, W., and Mayer, V. (1992). Biological and Earth Systems Science: A program for the future. *The Science Teacher*, 59, 32-37.
Kawasaki, K. (1996). The concepts of science in Japanese and Western education. *Science and Education*, 5, 1-30.
Lamy, S. L. (1990). Global education: A conflict of images. In. K. E. Tye, *Global education: From*

thought to action, (pp. 49-63). Alexandria, VA: 1991 ASCD Yearbook, ASCD.

Mayer, V. J., (1989). Earth appreciation. *The Science Teacher*, 56, 3, 60-63.

Mayer, V. J., (1991). Earth Systems Science: A planetary perspective. *The Science Teacher*, 58, 1, 34-39.

Mayer, V. J., (1995). Using the Earth system for integrating the science curriculum. *Science Education*, 79, 4, 375-391.

Mayer, V. J., (1997a). Earth Systems Education: A case study of a globally oriented science education program. In Merry M. Merryfield, Elaine Jarchow, and Sarah Pickert, Editors, *Preparing teachers to teach global perspectives*. Thousand Oaks, CA: Corwin Press, Inc.

Mayer, V. J., (1997b). Science literacy in a global era. *Hyogo University of Teacher Education Journal*, 17, 3, 75-89.

Mayer, V. J., and Armstrong, R. E. (1990). What every 17-year old should know about planet Earth: The report of a conference of educators and geoscientists. *Science Education*, 74, 155-165.

Mayer, V. J., Fortner, R. W., and Hoyt, W. H. (1995). Using cooperative learning as a structure for Earth Systems Education workshops. *Journal of Geological Education*, 43: 395-400.

Mayer, V. J., and Fortner, R. W. (Eds.) (1995). *Science is a study of Earth: A resource guide for science curriculum restructure*. Columbus, OH: The Ohio State University Research Foundation.

Millar, R. and Osborne, J. (Eds.) (1998). *Beyond 2000: Science education for the future*. London: Kings College.

Ministry of Education, Science, and Culture (1997). *Monbusho*. Tokyo: Ministry of Education, Science, and Culture.

Nakamura, H. (1980). The idea of nature, East and West. In Mortimer J. Adler, Editor in Chief, *The great ideas today, 1980*, (pp. 234-304). Chicago: Encyclopedia Britannica, Inc.

National Research Council (1993). *Solid-Earth sciences and society*. Washington, DC: National Academy Press.

National Research Council (1996). *National science education standards*. Washington: National Academy Press.

Orion, N. (1998). Earth sciences education + environmental education = Earth systems education. In Fortner, R. W. and V.J. Mayer, (Eds), *Learning about the Earth as a system. Proceedings of the Second International Conference on Geoscience Education (pp.* 134-137*)*. Columbus, OH: Earth Systems Education, The Ohio State University. ERIC Document ED422163.

Smith, A. F. (1994). A brief history of pre-collegiate global and international studies education. In J. Fonte and A. Ryerson, (Eds.), *Education for America's role in world affairs*, (pp. 1-22). New York: University Press of America.

Tye, K. E. (1990). Introduction: The world at a crossroads. In K. E. Tye, (Ed.), *Global education: From thought to action*, (pp. 1-9). Alexandria, VA: 1991 ASCD Yearbook, ASCD.

CHAPTER 2: A CASE HISTORY OF SCIENCE AND SCIENCE EDUCATION POLICIES

Victor J. Mayer
and
Rosanne W. Fortner
The Ohio State University, USA

1. INTRODUCTION

In this chapter we examine the apparent link between the history of national science priorities and the nature of the science curriculum in one country, the United States of America. We suspect that equivalent links can be found in most other countries, especially those that have aspired to some form of international leadership in politics and commerce. We document here how national priorities in the United States and the resulting political structure of the science establishment over the past century have resulted in a representation of the nature of science in school science curriculum that is inconsistent with the challenges facing the science establishment in the post Cold War world. Especially influential has been the need to develop a source of science and engineering man power and the technology essential for maintaining a strong national defense and an economically competitive business community. Science curricula are heavily influenced by the nature of the physical sciences since they have been successful in providing the scientific foundation for establishing and maintaining a powerful military and industrial/commercial capability.

Stephen Gould (1986), Agassiz Professor of Zoology at Harvard University, describes our science education and its methodological emphasis as follows:

> Most children first meet science in their formal education by learning about a powerful mode of reasoning called "the scientific method." Beyond a few platitudes about objectivity and willingness to change one's mind, students learn a restricted stereotype about observation, simplification to tease apart controlling variables, crucial experiment, and prediction with repetition as a test.

He goes on to point out that science curricula fail to provide a background in an essential component of the system sciences, that of history. In fact, they condition students to feel that a science that focuses on description and one in which experiments cannot be conducted is not science at all.

> These classic "billiard ball" modes of simple physical systems grant no uniqueness to time and object--indeed, they remove any special character as a confusing variable--lest

25

V.J. Mayer (ed.), Global Science Literacy, 25–35.
© 2002 *Kluwer Academic Publishers. Printed in the Netherlands.*

repeatability under common conditions be compromised. Thus, when students later confront history, where complex events occur but once in detailed glory, they can only conclude that such a subject must be less than science. And when they approach taxonomic diversity, or phylogenetic history, or biogeography--where experiment and repetition have limited application to systems in toto--they can only conclude that something beneath science, something merely "descriptive," lies before them.

Gould effectively portrays the type of science presented in American classrooms, its character developed through a century when our national focus has been on war and economic competition. He describes, in essence, a science curriculum modeled on physics and largely ignoring the contributions to science of both the methods and knowledge of the system sciences.

In this chapter we briefly discuss 'reduction' science and 'system' science methods. In Chapter Three we provide a more extensive discussion of the relative nature of these two aspects of science methodology as ends of a continuum of methods of science. We use the term 'reduction' when referring to the methodology of physics, chemistry and some of biology. It is a simple and descriptive term to use in characterizing the science methodology and the resulting product of these sciences. It informs on two levels, first methodological. Practitioners of these sciences seek to isolate single variables and test one against another, the 'billiard ball' analogy of Gould. In other words, they attempt to reduce the number of variables to be tested so that controlled experiments can be conducted. On another level, that of the information provided by the science, it is also descriptive. Physics especially, seeks to reduce the complexity of the world down to its simplest and most powerful elements such as time, gravity, nuclear attraction. In contrast the system sciences, which include the Earth sciences and ecology, attempt to study a system whole and over time. Thus, in Gould's terminology the system sciences are characterized by history and diversity of variables. Description and sequencing phenomena in time are fundamental. Science methodology is complex and at times each of the science disciplines calls on a variety of methods to elucidate an object or phenomenon, from reduction methods at one end of a continuum of methods to the system methods at the other end

Through the Cold War epoch especially, school science curricula, reflecting a national effort to recruit talented people into science, have emphasized the use of the investigative methods of chemistry and physics and the understanding of the natural laws and principles these sciences have developed. The curricula seldom make reference to the Earth system in which these laws and principles function nor do they include the science methodology of the system sciences. This has helped to create in the public mind a hierarchy in science with physics at the top thus contributing to a social and political climate in the USA where physics is seen as the embodiment of all science or by some, the only science. It is also an attitude often expressed in the physics community itself. Perhaps this attitude is best represented by a statement attributed to Ernest Rutherford (1871-1937) a prominent physicist, "All science is either physics or stamp collecting." Although Rutherford was English and did his research in England and Canada early in the last century, the attitude toward other sciences as expressed in this statement can often be found in the American physics community today.

Obviously not all science offered in the secondary schools is either physics or chemistry. Social and health concerns have provided a prominent place in school curricula for biological science concepts. Increasingly however, biology is being represented by the reduction approaches and molecular content rather then the ecological or systems approaches and content. Minimized or ignored in school science curricula is the study of how basic physical, chemical and biological processes act within their natural domains, the Earth systems of which they are functioning parts, and the science methods that can effectively study these processes. Adapting the science curriculum to represent the science methodology and conceptual contributions of all sciences, therefore, requires the inclusion of science methodology and content that goes beyond the approaches and content of the physicist and chemist. Instead the future students of science should also examine large systems as they normally function in nature. They need to learn how scientists collect and analyze observational data now aided by satellite and computer technology. They should learn how to use some of that data themselves in developing scientific explanations of natural phenomena just as they examine physical principles in laboratory situations. Only by participating in this type of 'system science' will they be prepared as citizens to evaluate its results and its power in informing them of the current status of our planet and potential solutions to environmental and social problems. Curricula that include systems oriented science can correct the current imbalance in secondary science programs and demonstrate to students how basic processes operate within systems and how systems are changed through human interventions.

After all of the money and time that has been invested in science curriculum restructure over the past decade, why is the result still deficient in the system sciences? Here we briefly review the development of science in the USA from the end of the Civil War to the present in support of our contention that national priorities shape a politics of science which in turn influences not only what science is practiced but also how science is represented in school curricula. We also take a brief look at the development of science curricula, which seems on the surface at least, to bear out our contention that the politics of science also influences the focus and content of science curricula. We then discuss briefly the neglect of system vis-a-vis reduction science. A more detailed examination of the philosophical basis for our arguments has been made in Chapter Three where we also relate system science to Eastern thought and suggest that it is complementary to certain Asian cultures. It therefore reflects a science more international in scope and culture than that currently in science curricula in most countries. We argue for changes that we believe are consistent with the directions being taken by modern national and international priorities for science and the emerging new politics of science.

2. A SHORT POLITICAL HISTORY OF AMERICAN SCIENCE

A discussion of the political history of science in the United States over the past century illustrates the effects of national priorities, often for defense, on our science establishment and the public's opinions regarding science. The era following the

Civil War in the USA was the time during which the foundation of American science was laid. The late 1800s was a crucial period in the Industrial Revolution in the USA. There was a need for natural resources to support the developing industrial base for what would become one of the world's most vibrant economies. Kelves (1987) in his history of American physics points out that it was the Earth scientists who were at the apex of political and scientific influence during this period. Clarence King was perhaps the preeminent American scientist of his time. He, John Wesley Powell and Charles Walcott, are a few of the Earth scientists who were the science power brokers of the latter half of the 19th century and on into the early 20th century. King conducted the Geological and Geographical Exploration of the Fortieth Parallel (Wilkins, 1988). Powell was especially effective in raising the federal funds necessary for these explorations. He was responsible for several other Western expeditions, and also the political maneuvering that resulted in the establishment of the United States Geological Survey (Stegner, 1954). The nature of the science conducted was descriptive and historical with little or none of the reduction later to typify American science. Kelves (1987) compares the two sciences of geology and physics as they coexisted during the late 1800s. The Geological Survey had won the prestigious Cuvier Medal awarded by the French Academy of Science for its collective work. Geology in the United States in general had the respect of the Europeans for the quality of its science whereas physics was poorly done in the United States. There were more geologists then physicists and a higher proportion of the geologists were theorists (p. 37).

Kelves (1987) points out that this preeminence of geology was in part the result of the considerable political influence exerted by Powell and other Earth scientists to secure the governmental funding necessary for the growth and institutionalizing of their science (p. 49-55). This achievement was influenced by the priorities of a country deep into the Industrial Revolution and in need of the natural resources of its domain even as later changes in the science establishment were influenced by war and its imperatives. Further evidence of the relative status of the Earth sciences in the scientific establishment of the late 1800s and early 1900s is found in the leadership of the National Academy of Sciences and the American Association for the Advancement of Science (Stegner, 1955; Wilkins, 1988).

According to Kelves (1987), the changes in the American science establishment began during the later stages of World War I when the requirements for winning that war became a priority for our political and industrial establishments. The Bureau of Mines was awarded substantial money by the Army to supervise more than seven hundred chemists working on chemical weapons (p. 132). ". . . observers then and since have, with considerable justification, called World War I 'a chemist's war' " (p. 137).

With the sinking of the Lusitania came an urgent concern, the growing German submarine threats to American shipping. With the failure of Thomas Edison's inventive genius to find a way to locate enemy submarines, the War Department called upon the academic physicists for help. They provided the conceptual knowledge of sound waves and their transmission through water which they used in the successful development of sonar (Kelves, 1987, p. 121-131). Subsequently, the federal support of science shifted to physicists and chemists with their sciences consequently becoming the major power centers in American science. With the onset of the Second World War more and more support flowed to the physical scientists and along with the money went prestige and political influence. Kelves states that, ". . . this was a war of physicists" (p. 320). The Los Alamos generation, brought to the forefront by its development of the Hiroshima and Nagasaki bombs, " . . . was dominated by physicists who seemed to wear the 'tunic of Superman,' in the phrase of a *Life* reporter, and stood in the spotlight of a thousand suns" (p. 334).

The ascendancy of the physical scientists in the mind of the public was matched by their ascendancy to the top of the hierarchy of science. The physical sciences, physics and chemistry and their science methodology best served the needs of the political and industrial establishments, advised by the same physical scientists, into and through the Cold War period. Thus it is Feynman, Teller, Alvarez and their colleagues and successors who have had the political influence to affect the direction of science. (Alvarez, 1987; Panofsky, 1999). Their success in assisting in the winning of the Second World War for the Allies and their work on nuclear physics during the Cold War placed their methodology at the helm of good science.

3. CHANGES IN SECONDARY SCHOOL SCIENCE CURRICULA

A rough parallel can be drawn in the evolution of secondary school science curricula over the same time period. In tracking the fate of the system sciences in the nation's curricula we use that of the Earth sciences. In the future more detailed research might also focus on the fate of ecology as a component of biology curricula. The Committee of Ten of the National Education Association, in 1893, presented its recommendations for the nature and content of science courses to be offered at the senior high school level. Its report, along with others directed at establishing requirements for college entrance, heavily influenced high school administrators and teachers in the development of their science courses (DeBoer, 1991, p. 40-50). These developments eventually lead to a sequence of science courses commonly offered in high school. They lead off with physical geography at the 9th grade, followed by biology in the 10th, chemistry in the 11th and physics in the 12th. This sequence has been labeled the 'layer cake' approach and, except for physical geography, has become the traditional sequence of offerings at the secondary level in the USA.

A topic outline for the physical geography course of the turn of the century would outwardly bear a very strong resemblance to that of a modern 9th grade Earth science course heavily oriented toward geology and geomorphology topics. In 1900 more than 20 percent of the high school population was studying physical geography, becoming the most frequent science course completed by high school graduates. Following World War One it gradually disappeared from high school curricula. In part it was replaced by general science under the influence of the progressive education movement. During World War Two, biology came to dominate as a 9th grade offering for more talented students. By 1950 most schools had eliminated physical geography from the curriculum, with the exception of New

York State, where physical geography became Earth science and has continued to be the 9th grade science Regents offering (Mayer, 1986).

In the 1960s, the Earth Science Curriculum Project (ESCP) provided curricular materials in support of a resurgence of interest in Earth science offerings at the 9th grade. This was in large part, an effort to provide higher quality 9th grade science programs replacing general science and to prepare students to take biology, chemistry and physics in later years. By the 1972-73 school year this resurgence was apparent as roughly 1,200,000 students enrolled in Earth science courses in grades 9-12 whereas about 1,000,000 were enrolled in general science courses (Mayer, 1978). This was perhaps the high water mark for enrollment in Earth science courses.

Science Indicators of 1995 (NSF, 1996), in reporting growth in high school science enrollments, documents increasing enrollments in biology in the period from 1982 to 1992 from roughly 75% of high school students to almost 90%, in chemistry from 30% to about 58% and in physics from roughly 15% to about 22%. Earth science enrollments were not even mentioned and can thus be assumed to be negligible at the 9th-12th grade levels (p. 39). Thus after a brief spurt of interest in Earth science in the early 1970s, spurred on by national funding of several Earth science curriculum programs and the American space program, interest fell dramatically. The lack of attention being paid to the Earth sciences in science curricula following the demise of physical geography in the early 1900s is starkly illustrated in DeBoer's (1991) discussion of the history of science curricula in American schools. Simply examining the book's index indicates the relegation of the Earth sciences to obscurity and irrelevance, at least in DeBoer's analysis. Physics has no fewer then 21 separate references, chemistry, 17, and biology, 21. Where does Earth science fall in its influence on science curriculum? It has three lines of reference. These refer to the Committee of Ten recommendations and to a simple one line reference each to the ESCP and the Time, Space and Matter curricula.

Thus we conclude that the history of emphasis of science curricula in the USA roughly parallels the political development of science from the late 1800s to the present. We are not implying that there is a cause and effect relationship, only that the facts are supportive of such a relationship. Certainly there are other possibilities for the demise of physical geography and Earth science in school curricula. DeBoer (1991), for example, suggested that the abstractness of the physical geography course and its declining popularity in the early 1900s (pp. 86-87) caused its displacement by general science. But it appears that general science was no more popular than that which it replaced, nor were the other science courses that remained largely unchanged very popular.

The centrality of the physical sciences to the Cold War effort had its influence on the content of science curricula. The efforts to improve science education beginning in the early 1950's were lead by physicists such as Zacharias of Massachusetts Institute of Technology. The nature of the science curriculum largely reflects the content and operating procedures of the physical sciences. Even the Earth Science Curriculum Project (ESCP) materials emphasized the use of controlled experiments partly justifying the curriculum's use in preparing students for later enrollment in chemistry and physics. One experiment, for example, had students shake cans of rock in a controlled experiment to determine erosion rates (ESCP, 1967, p. 19).

President Eisenhower's Economic Report for Fiscal Year 1957 which underlined the defense needs for quality science education in secondary schools excludes the 'system sciences'. The report draws attention to the need for better high school instruction in science:

> The supply of scientific and engineering manpower, so vital to our military security and future prosperity, depends to a large degree on the availability and quality of such instruction. Physics and chemistry are not taught at all in many high schools, and in many others the teaching of these subjects is seriously deficient. (Krieghbaum and Rawson, 1969, p. 186)

No mention is made of biology or the Earth sciences.

Following the launching of Sputnik by the Russians in the fall of 1957, a critical examination of our educational system was begun, with the major emphasis placed upon science education. There was widespread agreement within the political establishment that the future military and economic superiority of our country depended upon maintaining its lead in science. As a result, the federal government committed large sums of money to the improvement of science education. The National Science Foundation granted millions of dollars to several organizations to develop curricula in physics and chemistry and to improve the science background of secondary school science teachers. This was followed in the 1960s by projects in biology and the Earth sciences including the Earth Science Curriculum Project (Krieghbaum and Rawson, 1969). Duschl (1990) draws attention to the sequence in which the various curriculum projects were funded with physics first followed by chemistry and then biology--the reverse of the sequence normally taught in American high schools. This sequence of funding set as a primary goal for the restructuring of the American science curriculum, "students' attainment of the highly mathematical and logical final form version of science found in physics and chemistry" (p. 24).

4. FOCUS OF CURRENT AMERICAN CURRICULUM REVISION EFFORTS

During the Reagan Administration American education, and especially science education, came under severe criticism once again for failing to stem an economic decline (National Commission on Excellence in Education, 1983). As a result, there were a number of science curriculum renewal efforts. In a review of eight, all developed in the late 1980s, Raizen (1997) summarized the factors contributing to the support of the projects. They include "the need for the United States to maintain its competitive position in the world economy . . . " which was seen as increasingly dependent upon technology. The second factor was "the need for the average citizen to know and understand enough to deal in an informed way with . . . decisions increasingly linked to science and mathematics . . ." and the third ". . . the low standing of U.S. students in a number of international assessments" (p. 21-22).

The most prominent of the eight projects, all focused on typical Cold War objectives for science, was Project 2061 of the American Association for the Advancement of Science (AAAS). It was conceived and directed under the leadership of James Rutherford, a former high school physics teacher in California, professor of science education at Harvard, and director of Harvard Project Physics, a 1970s National Science Foundation supported curriculum development project. Started in 1985, the purpose of the project was to "define the knowledge, skills, and attitudes all students should acquire" regarding the science enterprise. The five panels of scientists, mathematicians and engineers charged with the definition conducted their work during what later came to be seen as the closing stages of the Cold War epoch, from 1985 until 1988. The 49 scientists serving on the panels were broadly representative of the sciences, engineering and mathematics (AAAS, 1989, p. 234-236). However, from titles alone we could only identify one marine biologist on the Biological and Health Sciences panel, one ecologist on the Technology panel, one geologist and one geographer on the Physical and Information Sciences and Engineering panel. There were several individuals identified simply as Professor of Biology on the Biological and Health Sciences panel and they could very well have a specialty in ecology. The panels were very heavily weighted toward engineering, mathematical and computer sciences and had very minimal representation from the systems sciences. The report of these panels, submitted to the AAAS membership, was quite controversial. The chemists and geologists especially raised formal concerns about a perceived slighting of their sciences.

The summary of the five working panels written by Rutherford and his colleague Andrew Ahlgren was embodied in *Science for All Americans* (AAAS, 1989). There is no reference in this final report to the concepts behind Earth System Science (Earth Systems Sciences Committee, 1988) now perhaps the major thought system revolutionizing the Earth sciences (see Chapter One). This is not surprising, since not a single member of the Earth Systems Science Committee (ESSC) was represented on any of the five panels. Project 2061 and its products seem to represent the vision of those scientific communities with the most financial support and political influence during the Cold War epoch. As indicated in Raizen's (1997) review, it reflects the use of science as a basis for the development of technology for defensive and offensive purposes, both military and commercial.

5. A TIME FOR CHANGE IN EMPHASIS FOR SCIENCE

President Dwight Eisenhower, when he left office in 1961, warned the citizens of the United States to be alert to the potential for dominance by the 'military-industrial complex' and the scholars enlisted in its support:

> "In the councils of government, we must guard against the acquisition of unwarranted influence . . . by the military-industrial complex." We must "gravely" regard the "prospect of domination of the nation's scholars" by federal largess. "We must also be alert to the equal and opposite danger that the public policy could itself become the captive of a scientific-technological elite." (Quoted by Kelves, p. 393)

Despite his warning, more and more money went into military and nuclear research programs during the Cold War period and consequently the physics research community. In turn, the science curriculum in the secondary schools of the USA and through its influence, in many countries around the world, came to focus on the type of science supportive of Cold War technology, that of the physical sciences. Now, however, it is time for change in the nature and objectives of science and therefore the science curriculum to reflect new emphases.

The Honorable George E. Brown, Jr., until 1999 Representative in the United States Congress from California and a longtime leader in science legislation, stated at the Fellows Reception held by the AAAS in 1994 that for the science enterprise to remain relevant to national priorities and effectively contribute to the solution of the major post-Cold War problems experienced by many nations it must refocus its priorities, capabilities and activities. Brown, quoted later in a newspaper article, criticized the results of our current science and technology development (in Lepowski, 1994):

> Today there are more human beings living in abject poverty throughout the world than ever before. At home, our global leadership in science and technology has not translated into leadership in infant health, life expectancy, rates of literacy, equality of opportunity, productivity of our workers, or efficiency of resource consumption.

Brown goes on to question the future role of science in our society:

> Must science and technology continue to feed the historical cycle of more consumption, more waste, more economic disparity? Or can our research lead us out of that cycle, and create a new trajectory for cultural evolution?

Thus Brown argues for a different direction for science, one that will address the problems of environmental degradation and social upheaval. For science to continue to be seen as relevant to the needs of our country and therefore continue to receive the moral and financial support of its citizens, it must be able to help solve these problems, not continue on the old trajectory established during the Cold War.

Lubchenco (1998), in her presidential address to the AAAS membership also reflected on the changing needs for science:

> Fundamental research is more relevant and needed than ever before . . . Just as the Manhattan Project involved a major investment in fundamental research, adequately addressing broadly defined environmental and social needs will require substantial basic research . . . We can no longer afford to have the environment be accorded marginal status on our agendas. The environment is not a marginal issue, it is the issue of the future, and the future is here now.

In her talk presented four years after Brown's discussion, Lubchenco reinforces the need for a change in direction of science to meet environmental and social concerns.

As Brown and Lubchenco have stated, national priorities for science must change to confront the social and environmental problems brought on in large part by the side effects of national policies in science and technology during the past century. We believe that educators should be leaders in redefining the nature of the pre college science curriculum and in this way have a hand in assisting the reshaping of the political structure and objectives of the science establishment. To do this, however, we must have a clear and broad understanding of the nature of science, its history and its current structure and priorities.

6. CONCLUSION

The knowledge derived through the system sciences has had far-reaching impact on our intellectual and cultural lives. Historically, these sciences have dramatically changed our perception of our place in nature, by Copernicus and Galileo redefining the Universe with the Sun at the center orbited by the Earth, by Hutton in developing the concept now popularly referred to as 'deep time', by Darwin as he formulated his theory of natural selection. These sciences impact our lives far beyond that suggested by Lederman and his panel (Atkin, et al, 1998) who stated "the influence of science on society is mainly through technology." Although technology has become a major component of the relationship between science and society, it is the science constructs of Copernicus, Hutton, Darwin and others that provide us with a philosophical place for human existence. These are powerful contributions of science to our society. In addition science can lay the foundation for an aesthetic appreciation of our world and its beauty, and of its long and complex history. Our citizens also need to be knowledgeable about the system science methodologies and their role in assessing changes in our global environment, just as they are now made aware of reduction methods and their contributions to our rapidly developing technology. Thus, science curriculum specialists need to include the system sciences and their products. Indeed the knowledge provided by the system sciences can provide the conceptual structure for organizing science curricula.

We have tried to make the case that the type of science conducted within the USA is the result of priorities outside of science itself. They originate in the political system as it responds to the needs of our nation. The nature of the science curriculum follows these priorities. This is as it should be. Science must serve the needs of the nation. The political priorities of the past have produced a science curriculum focused on the nature of the physical sciences, neglecting the importance of the system sciences. In the new millennium, the nation's priorities for science are changing. Science must be redirected to assist in solving or ameliorating environmental and social problems resulting, at least in part, from the past use of science. As the nation's priorities for science change, so too should the science curriculum change. How can curriculum specialists restructure the science curriculum to accommodate these changes and develop a science curriculum that looks to the future rather then to the past? As we have discussed in Chapter One, Global Science Literacy, with its science foundation based in Earth systems science, can provide a conceptual and methodological structure for future science curricula, not only for American schools but for those of all schools in deomocracies worldwide.

REFERENCES

Alvarez, W. (1987). *Alvarez: Adventures of a physicist*. New York: Basic Books, Inc.

American Association for the Advancement of Science. (1989). *Science for all Americans*. Washington, D.C.: AAAS.

Atkin, J. M, Black, P., Lederman, L., Ogawa, M., Prime, G., and Rennie, L. J. (1998). FORUM: The ICSU Programme on capacity building in science, *Studies in Science Education*, 31, 71-136.

DeBoer, G. E. (1991). *A history of ideas in science education*. New York: Teachers College Press.

Duschl, R. A. (1990). *Restructuring science education: The importance of theories and their development*. New York: Teachers College Press.

Earth Science Curriculum Project. (1967). *Investigating the Earth*. Boston: Houghton Mifflin Company.

Earth System Science Committee. (1988). *Earth system science: A program for global change*. Washington: National Aeronautics and Space Administration.

Gould, S. J. (1986). Evolution and the triumph of homology, or why history matters. *American Scientist*, 74, 60-69.

Kelves, D. J. (1987). *The Physicists: The history of a scientific community in modern America*. Cambridge, Massachusetts: Harvard University Press.

Krieghbaum, H. and Rawson, H. (1969). *An investment in knowledge*. New York: New York University Press.

Lepowski, W. (1994). Science-technology policy: New directions in the Clinton era.. In *Science and technology policy yearbook, 1993*. Washington, D.C.: American Association for the Advancement of Science, pp. 109-120.

Lubchenco, J. (1998). Entering the century of the environment: A new social contract for science. *Science*, 279, 491-497.

Mayer, V. J. (1978). Enrollment in secondary school Earth science courses. *Journal of Geological Education*, 26,16-17.

Mayer, V. J. (1986). Earth science in American schools. *Geology Teaching* (Journal of the Association of Teachers of Geology, United Kingdom), 10, 4, 20-21.

National Commission on Excellence in Education (NCEE). (1983). *A nation at crisis*. Washington, DC.: United States Department of Education.

National Science Foundation. (1996). *Indicators of science and mathematics education, 1995*. Washington, DC: National Science Foundation

Panofsky, W. K. H. (1999). Physics and government. *Physics Today*, 52, 3, 35-41.

Raizen, S. A. (1997). The general context for reform. Chapter 2 in Senta A. Raizen and Edward D. Britton, Eds. *Bold ventures*. Boston: Kluwer Academic Publishers.

Stegner, W. E. (1954). *Beyond the hundredth meridian: John Wesley Powell and the second opening of the West*. Boston: Houghton Mifflin.

Wilkins, T. (with the help of Caroline Lawson Hinkley). (1988). *Clarence King: A biography*. Albuquerque: University of New Mexico Press.

CHAPTER 3: THE PHILOSOPHY OF SCIENCE AND GLOBAL SCIENCE LITERACY

Victor J. Mayer, The Ohio State University, USA
and
Yoshisuke Kumano, Shizuoka University, JAPAN

This chapter is an adaptation of an article originally published in *Studies in Science Education*, vol. 34, pp. 71-90 (1999) under the title of: The Role of System Science in Future School Science Curricula

1. INTRODUCTION

There has been a significant amount of discussion in the science education literature concerning needed changes in the science curriculum. These discussions have centered on several proposed reasons for change. One is based on the relative performance of a nation's children on international tests, the major political driving force behind the efforts of the American Association for the Advancement of Science (AAAS) and the National Academy of Sciences (NAS) in the United States to produce guidelines (AAAS, 1989; NAS, 1996) for the improvement of science curricula targeted as contributing to these low test performances (Bracey, 1998; Wang, 1998).

Another argument comes from science educators engaged in multi-cultural education in the USA and some science educators in non-Western countries and is based partly on low student performances in science and low representation of women and American minorities in scientific careers (Klotz, 1993; Ogawa, 1995; Stanley and Brickhouse, 1994). Black and Atkin (1996, p. 16-17) in a study of curricula from 13 countries sponsored by the Organization for Economic Co-operation and Development (OCED) identified a concern with inclusiveness and equity as a driving force for curricular reform in science, mathematics and technology. The Ministry of Education, Science and Culture (Monbusho) in Japan, has expressed concern regarding the nature of the Japanese curriculum generally and its role in developing more creative, independent and problem solving characteristics among Japan's citizens (Monbusho, 1997, p. 9).

A major factor is the concern among scientists and science educators of several countries, including Great Britain, of the need for science curricula directed toward developing the understanding among the general citizenry of the important processes and concepts of science (Millar and Osborne, 1998; Black and Atkin, 1996). This contrasts with curricula developed in the USA in the 1960s and 1970s and other countries as well that seem to be designed to attract students into the science and technology professions. This more recent focus upon scientific literacy has also been the guiding philosophy of the efforts of the AAAS and the NAS in the development of curriculum guidelines for schools in the USA.

V.J. Mayer (ed.), Global Science Literacy, 37–49.
© 2002 *Kluwer Academic Publishers. Printed in the Netherlands.*

Although these discussions and contributions are of interest and importance in academic dialogue and curriculum restructure, they do not get to the core of what we believe to be a major challenge for the school science curricula of the future. What is apparently absent from current restructure efforts is any substantive analysis of the future needs for science in the post-Cold War era, how it must, and is, adjusting to these new priorities, and the implications for an adequate representation of science in the world's pre-college science curricula. We do see a substantial reference to technology and its relation to science and the importance of its inclusion in science curricula. The argument made for its inclusion is usually to demonstrate the relevance of science to students (AAAS, 1989; Atkin, 1998; Millar and Osborne, 1998; Monbusho, 1997). This, however, can also be seen as carrying on a role for pre-college science in preparing and recruiting students into science and technology careers in support of international commercial competition, one of the driving motives for curriculum restructure found in the OCED study (Black and Atkin, 1996, pp. 14-16).

In reviewing all 23 programs for common factors influencing curricular reform, Black and Atkin (p. 60) also conclude that the role of academic scholars in defining the 'basics' of science and mathematics from research results and their importance in curricula for all students was found in only one program. That was in Project 2061, completed and published in 1989, toward the end of the Cold War. It drew upon a science, mathematics and engineering establishment however still embedded in the science politics of war and economic competition.

2. POST-COLD WAR CHALLENGES TO SCIENCE AND SCIENCE EDUCATION

The end of the Cold War and the beginning of the era of globalization are factors that should bring substantial reexamination of the historical and philosophical foundations of science upon which current science curricula have been established. Such a reexamination would have implications for the content and focus of future curriculum efforts whether directed at developing curricula to attract and prepare future scientists and engineers, or for the general education of a citizenry in science. Supporting this contention is the fact that some within the scientific and political communities are working to refocus the objectives of a science establishment employed for much of the last century in developing the scientific foundation for the conduct of war (Kelves, 1987) and commercial competition perhaps starting as far back as the Industrial Revolution in Britain. As the goals and objectives of the science establishment are reshaped, so must science curricula be changed to reflect the modern priorities of science.

With the scientific documentation of the environmental and social problems infecting developed and developing nations, and indeed the Earth system, many scientists and political leaders feel that science must be redirected toward the solution and prevention of these problems. In Chapter Two we have discussed the viewpoints of George E. Brown, Jr., representative in the United States Congress from California until 1999, and a longtime leader in science legislation, and Lubchenco (1998), in her presidential address to the AAAS membership. Both have concluded that the priorities for science must change so that it can more effectively confront the social and environmental problems brought on in large part by the side effects of policies in science and technology during the past century. Educators should be leaders in redefining the nature of the pre-college science curriculum and in this way participate in reshaping the political structure and objectives of the science establishment. To do this, however, we must have a clearer and broader understanding of the nature of science, its history and its current structure.

3. THE HISTORY AND PHILOSOPHY OF SCIENCE

Matthews (1994) believes that there are important lessons to be learned by science teachers from the history and philosophy of science. Such knowledge will be essential for the improvement of science teaching. We endorse his position with a warning to be alert concerning the nature and perspective of much that has been written about the history and philosophy of science. Almost invariably, those who study and write about it focus their attention on the physical sciences, physics and chemistry. This provides a view of the world that is mechanistic and deterministic, and of a method of science in which scientific constructions are developed of components isolated from their natural systems. Much of what is written about the history and philosophy of science is in reality the history and philosophy of physics. Arthur Strahler (1992) in his introductory chapter on the nature of science states it thus:

> One difficulty is that the giants in science and philosophy are prone to restrict the scope of science rather severely, usually to physical science, and even as narrowly as to nuclear and quantum physics alone. We must on the other hand, cover a large area of natural science quite far removed from the ideal behavior of matter on a subatomic scale. Besides dealing with the origin and physical evolution of the universe--a field that does indeed rest in large part on principles of theoretical physics--we must include the geological and biological evolution of our own planet Earth over a time span of billions of years. Here we will find extremely complex aggregations of matter that have long and involved histories of continuous development. Scientists who investigate these historical areas of knowledge need to adopt specialized views of science. (p. 7)

Seldom do writers on the philosophy of science draw from the historic and interpretive sciences of geology or ecology. An exception, in addition to Strahler, is Robert Frodeman (1995) a researcher in the philosophy of science who possesses a graduate degree in geology. As evidence of the neglect of what he terms the 'interpretive and narrative' forms of science he points out that while philosophers of science widely acknowledge the crucial importance of the Copernican Revolution in our conception of space, they ignore the even more significant contributions of Hutton and Werner in the development of the concept of the great age of the Earth, geological time or what is popularly referred to as 'deep time'.

Strahler is a geologist who has written a book for science majors entitled *Understanding science.* He categorizes areas of knowledge as belief, such as religion and political ideology, and the empirical sciences. The empirical sciences, through research, construct reliable but not infallible knowledge of the real world and include explanations of the phenomena (p. 8). Empirical science differs from belief in that it " . . . must follow rules of sound logic and must make use of mathematics" (p. 13).

Strahler, in order to clarify the scope of science more broadly than is typical in the writings of the physical science community, describes three fields of empirical science (p. 13): 1) the physical sciences which deal with the nature of matter and energy and largely utilize controlled experiments conducted within laboratories; 2) several complex inorganic sciences requiring observation of nature outside of laboratories such as oceanography and geology, and; 3) the biological sciences which deal with living cells and their aggregations. Unique to the second and third fields is their reliance upon the history of events spread over vast periods of time. Strahler and other Earth scientists use the term 'history' as the sequence of events that have occurred over a defined period of time, as in the 'history of dinosaurs', and the physical and biological processes that they mark. It should not be confused with and does not refer to the nature of investigation used by historians in interpreting human history.

In a later chapter, he goes into more detail in describing the various areas of the natural and social sciences. Under what he calls the pure sciences he lists the natural and social sciences, and under the natural sciences, the physical sciences (chemistry and physics), the macrocosmic/inorganic sciences (including geology and meteorology among others) and the biological sciences. He characterizes the physical sciences as atomistic, reductionistic, timeless, and universal. They are,

> ...set apart from the other natural sciences on grounds that they seek to understand the basic nature of matter and energy and their interactions; while the others study complex and unique systems of matter and energy that occupy specified and fixed positions in time and space.

> ...they are concerned with general (universal) laws that can be expressed mathematically or by other symbols. The method of the physical sciences is almost entirely experimental, including not only laboratory experimentation but also repeated instrumental observations in a natural environment. (p. 77)

The macrocosmic/inorganic sciences and the biological sciences (MIS/BS) are complex/historical, synthetic/emergent; organicistic, time bound, and particular.

> Besides dealing with highly complex structures, the MIS/BS investigate complex time sequences of events, something we do not find in basic physics and chemistry. Examples are the life cycle of an individual of a plant or animal species, the evolution of life on earth over hundreds of millions of years, or the birth and life history of a star or galaxy. Thus, the quality of history pervades MIS/BS. (pp. 80-81)

We have presented a condensation of his categorization of the natural sciences, not to suggest that there are different sciences, but to point out the narrowness of the traditional discussions of the history and philosophy of science in the science education literature. Such discussions normally rely on what Strahler categorizes as the physical sciences. Their narrowness, we believe, has also resulted in a narrowness of the representation of science in school curricula. As Duschl (1990, p. 24) has pointed out, the physical sciences have had an overriding effect on the nature of science curricula, not only in the United States and Japan, but worldwide. Yet, it is the MIS/BS sciences and their methodology that may make the difference in effectively facing the environmental and social problems that should become the major topics of scientific inquiry in the future.

3.1. The nature and importance of system science

Frodeman (1995) describes the method of scientific reasoning used by geologists. He makes the argument that this methodology is related to but significantly different from that used by the physicist. In part, this is because little of interest to geologists can be done in the laboratory under controlled conditions. Investigations go on in the natural environment with only minor confirming types of experiments that can be done in the confines of a laboratory. Geologists work in a historical science where descriptions of events occurring in the past need to be made by analogy with current processes. Many variables, some present temporally, others acting historically, must be considered when drawing conclusions. As a result, geologists use an interpretive and narrative form of reasoning. It is this type of reasoning, however, which he concludes will be of greatest value as science explores current and future environmental and social challenges. Primarily through this type of reasoning can we gain the knowledge to reduce the effects of and adapt to global warming, correct problems caused by water and air pollution, and solve some of the social problems in the post Cold War era.

  Most Earth scientists would agree with the nature and importance that Frodeman ascribes to the intellectual processes they use. For example, Walter Alvarez (1997) a geologist and son of Luis Alvarez, Nobel Laureate in Physics, compares the processes used by the physicists and Earth scientists in the preface of his book written for a general audience. In discussing the science that has led to the understanding of the great Cretaceous extinction he states:

> It is also the story of how geology and the other disciplines which study the Earth have emerged as fully mature sciences, distinguished by their inherently interdisciplinary nature, by the complexity of their subject matter, and by the obvious requirement to move from reductionistic to holistic science in order to achieve their central goal of understanding the Earth. Through the twentieth century, physics and chemistry, and recently molecular biology, have made enormous strides in understanding Nature by the analytical approach--by reducing problems to their fundamental components and studying these components in isolation. In the twenty-first century, science will be in a position to begin putting the pieces together, in order to seek a synthetic or holistic understanding of Nature. The Earth sciences are inherently synthetic and are therefore uniquely placed to lead this development. (pp. x-xi)

  Lazlow (1972), a philosopher, describes what he terms 'system science' in some detail. He starts by redefining for his purposes the usual categories of nature as discussed by philosophers and historians of science--inorganic, organic and social--as sub organic, organic and supra organic. This therefore implies levels of organization of nature (p. 30) instead of distinct compartments or categories of objects or processes and makes the discussion of nature more amenable to a systems approach. In sub organic he includes phenomena such as atoms which, although systems, are essentially closed systems not requiring exchanges of energy from outside for their continuance and maintenance. He then goes on to define the organic as open ended systems, requiring energy and material exchanges with the environment. Unfortunately in his concept of levels of organization, he excludes purely open physical systems such as those contained in the atmosphere or hydrosphere. Even more complex in his organization scheme are the supra organic entities such as human groups and the personality and social interactions comprising such systems. He concludes that organic and supra organic systems, and we would add complex physical systems, are much too complex for the human mind to understand from the traditional atomistic or reduction approaches of the physical scientist. Instead, natural systems must be viewed as wholes with irreducible properties. Starting with his first characteristic that "Natural systems are wholes with irreducible properties", he establishes four characteristics of natural systems. The second characteristic is a function or result of the fact that complex systems are open systems exchanging energy and materials with their environment and other systems. Thus, "Natural systems maintain themselves in a changing environment" (p. 34). This includes social structures which, under changing conditions, adjust and adapt ". . . . maintaining themselves in a dynamic steady state rather then one of inert equilibrium" (p. 46). In describing his third characteristic, he uses some unfortunate words. Because natural systems are open " . . . they create themselves in response to the challenge of the environment" (p. 46). By this, he simply seems to mean that systems change to meet new environmental conditions--not that natural systems change themselves in some type of conscious process. As an example of organic structures, he used the idea of organic evolution and the persistence of organismal changes that are adaptive, thus resulting in more complex organic individuals and populations over time. His fourth and final characteristic is that "Natural systems are coordinating interfaces of nature's hierarchy" (p. 67). Here he seems to mean that any system above the atomic level is composed of a collection of subsystems, each level incorporated into an integrated supra system. Thus, a person can be seen as cellular systems organized into organ systems organized into an individual. The individual, in turn, is a subsystem of a family system and so on. Moreover, not to be lost sight of is that the individual exists in a physical and organic habitat, again, and so on. Each subsystem interacts and coordinates itself with those at the same level and those below and above:

> In sum, nature, in the systems view, is a sphere of complex and delicate organization. Systems communicate with systems and jointly form super systems. Strands of order traverse the emerging hierarchy and take increasingly definite shape. Common characteristics are manifest in different forms on each of the many levels, with properties ranged in a continuous but irreducible sequence from level to level. The systems view of nature is one of harmony and natural balance. Progress is triggered from below without determination from above, and is thus both definite and open-ended. To be 'with it' one must adapt, and that means moving along. There is freedom in choosing one's paths of progress, yet this freedom is bounded by the limits of compatibility with the dynamic structure of the whole (pp. 74-75).

This description of nature stands in sharp contrast to the view of a mechanistic and determinist universe--the product of the reduction sciences. However, Lazlow provides us with little insight as to the methods to be used by scientists in studying natural systems as irreducible wholes. For this we must look to Frodeman (1995).

Frodeman characterizes the reasoning methods of the geological sciences as **hermeneutic** or interpretive, and **historical**. We suspect that these are characteristics shared by what Strahler grouped as the macrocosmic/inorganic sciences (including geology and meteorology among others) and the biological sciences (MIS/S), and what Alvarez referred to as 'holistic science'. For want of better guidance and relying on first occurrence, we apply Lazlow's (1972) term 'systems science' to those sciences that rely on what Frodeman calls the hermeneutic or interpretive and historical reasoning methods.

According to Frodeman, hermeneutics come from the efforts to reconcile contradictory statements found in historical documents to help discover original meanings. The geologist applies this type of reasoning to the various characteristics of an outcrop, "judging which characteristics or patterns in the rock are significant and which are not". It is apparent that in this process, preconceptions will be involved in making judgments. Thus, the data used by the geologist will be meaningless to the uninitiated until "the geologist introduces concepts for 'seeing' the rock". In a sense, this type of reasoning may be a tool in the development of all human knowledge, even in the 'objective' physical sciences (Eger, 1992). However, it is basic to the system sciences. There are three characteristics of hermeneutics that play fundamental roles in geological reasoning. They include the "hermeneutic circle--the fore restructures of understanding, and the historical nature of knowledge". The hermeneutic circle is a process of back and forth reasoning where earlier conceptions are used as new data are presented to continue to build a conceptual structure. Thus, "wholes at one level of analysis become parts at another". This is the means by which all understanding progresses. The second point, fore structure, relies upon the scientist's preconceptions and foresight. Preconceptions are the theoretical basis upon which the interpretation of data is approached and foresight the "presumed goal of our inquiry and our sense of what will count as an answer". The final characteristic includes the tools and sets of procedures or practices that are brought to the collection and processing of data. The nature of these tools and procedures also shapes the types of data acquired and their interpretation. This will also include the discussions and critiques of colleagues as interpretations are developed or discussed in the literature. As these interpretations are accumulated over time, "the body of scientific knowledge comes to have a strongly historical component".

The 'systems' nature of the geological sciences can be characterized in three points: "the limited role or relevance of laboratory experiments, resulting in geology's dependence on other types of reasoning; the problem of natural kinds (i.e., the question of defining the object of study within historical geology); and geology's nature as a narrative science". The physical sciences are distinct from the system sciences in that time is not a factor. By using controlled reproducible experiments, causal analysis results through rather simple reasoning processes not involving history. In the system sciences however, since controlled experiments are impossible to construct, history becomes crucial and a "different set of criteria for what counts as an explanation" must be used. One of these is the use of reasoning by analogy, especially the assumption of analogies between present processes and those presumed to have operated in the past. A second challenge is identifying and defining the object of study. Thus, different interpretations can be constructed as a result of different definitions of what at first appeared to be an identifiable object. Finally, "the historical sciences are distinguished by the decisive role of narrative logic in their explanations. Narrative logic is a type of understanding where details are made sense of in terms of the overall structure of a story".

The book by Alvarez (1997) mentioned earlier, is a very vivid and interesting documentation of the nature of the system science approach to science investigation. It describes the history of investigations that resulted in solving the mystery of the extinction of the dinosaurs and some 70 percent or more of all species extant at the end of the Cretaceous Period.

4. THE CULTURAL IMPLICATIONS OF EARTH SYSTEM SCIENCE

Science is not static. It evolves just as the demands placed upon it change. For example, as discussed in Chapter One, several American federal science agencies, starting in 1986, cooperated on an effort to reshape the nature of Earth science research funded by the national government. The Earth System Sciences Committee (ESSC), charged with outlining the principles of this reformulation, called the result of their work 'earth system science' in their report (ESSC, 1988). This report describes the concept of the Earth as a system and the role of each of the science disciplines in contributing to a holistic knowledge of the processes operating within that system. It provides a holistic view of science and of the Earth system, a view that includes people and their interactions as integral components of the Earth system. It is also the theme contained in the subsequent report entitled, *Solid-earth science and society* (NAS, 1993). It seems to us that this view of nature is very consistent with the Japanese perspective of the oneness of humans with the Earth system and as mutually interacting components (Nakamura, 1980). Thus science curricula based on the concept of the Earth as a system and incorporating the science methodology of the system sciences should more closely approach the cultural beliefs of Eastern countries than do the traditional physical science constructs imbedded in reduction science, currently the basis for science curricula worldwide.

Although our reasoning is based upon the evolution of the Earth sciences, this view seems to be consistent with current thinking among practitioners in other science communities. For example, there is an element in the physics community that is critical of its focus on reduction and supportive of a move toward a more systems oriented science. Fritjof Capra is widely published for his research in high-energy physics. Since the 1970s, he has focused his attention on the dramatic changes that have occurred in concepts and ideas in physics since the turn of the century and their potential for a profound change in worldview that they have brought about. In his book, *The Taos of physics*, he compares the new understandings about the nature of our universe favorably with the understandings held by Eastern mystics. Unfortunately, from our point of view, this leads him to a discussion and belief in Eastern mysticism as the direction that the new physics is leading scientists such as him. Nevertheless, some of the conclusions he draws concerning the directions of science today are very relevant to our thesis. In the *Afterward* to the third edition (1991) of his book, he proposes six criteria that define the new paradigm, which he calls a 'holistic world view'. This is a view of the world as an integrated whole rather than the accumulation of distinct parts, the traditional view of the physicist. His first criterion marks science as moving from the view that the whole is the sum of its parts, to the view that "while the properties of the parts certainly contribute to our understanding of the whole, at the same time the properties of the parts can only be fully understood through the dynamics of the whole. The whole is primary . . ." (Capra, 1991, p. 328). His second criterion places process as primary rather then the traditional emphasis upon structure. Thus, it is the history or trend of process that is the primary concern of science not the relationships between fundamental parts.

Capra's third criterion concerns the process of studying a phenomenon and the role of the individual scientist and his or her perceptions and their effects upon the outcome. This introduces an uncertainty and shifts our understanding of science from being objective to being 'epistemic'. Here he seems to demonstrate a lack of understanding of the Earth sciences and their reliance on the thinking processes involving hermeneutics discussed earlier. Instead, he tries to project the implications of the Heisenberg Uncertainty Principle into an inappropriate area. His fourth criterion, which he calls the most profound of all, is that science, instead of describing fundamental objects or structures will focus on relationships between objects and structures. He uses the metaphor of a network. With these four criteria in mind, "Nature is seen as an interconnected dynamic network of relationships that include the human observer as an integral component" (Capra, 1991, p. 333). Since it is impossible to study every component of a system to understand completely the whole, science is moving from 'truth' to providing "approximate descriptions" of reality, his fifth criterion. His final criterion for science is one of advocacy. In it, he contends that traditional science is based on "man's" need to control nature:

> I believe that human survival in the face of the threat of nuclear holocaust and the devastation of our natural environment will be possible only if we are able to radically change the methods and values underlying our science and technology. As my last criterion I advocate the shift from an attitude of domination and control of nature, including human beings, to one of cooperation and nonviolence. (p. 334)

Capra now sees physics not as the model for all sciences, but as a special case of the systems approach, one that concerns itself only with the nonliving systems (p. 339).

The current change of thinking among certain scientists from different disciplines is also made clear in a discussion of the practice of modern biology in these comments of Harold J. Morowitz (1989) from his book entitled, *Biology of a cosmological science*:

> In the *Meeting of East and West,* F. S. C. Northrop has pointed out that Eastern thought is much more devoted to a directly given continuum view of reality, while Western thought is much more centered on theoretic constructs, such as objects . . . Biology has developed within the Western object-centered atomistic view of reality which has come to fruition in modern molecular biology. More recently, much of biological thought has

been moving from the Western orientation to a differentiated continuum view that is
much closer to Eastern philosophy. (p. 45)

Morowitz makes clear that people are a part of natural systems and if we truly
understood this, we would not have ecological crises:

> ... The build-up and breakdown of molecular order are linked to the environment around
> us; the inflows of energy must come from outside ourselves, as we in turn must radiate
> energy to our surroundings. Unless there is balance in the surrounding world, the
> individual processes we have been describing cannot take place. This is a biological,
> thermodynamic statement of the real necessity for environmental balance.
> Environmental balance is not something that one must achieve for aesthetic reasons; it
> is the very key to transient existence in a flow-dominated nature. If we misdirect the
> flows to re-alter the system, the environment may undergo radical changes, and it may
> no longer be possible to maintain the kind of stability that we require. (p. 48)

In a sense, modern biology is experiencing duality. As the molecular approaches
now so productive of bioengineering developments predominate in certain research,
the understanding of the unity and coherence of nature is also becoming increasingly
sophisticated emphasizing the need for applying systems science methodology to
our understanding of the biosphere and its interactions with each of other Earth
systems. Both are essential to our understanding of biological systems. However, the
commercial promise of the molecular or reduction approaches results in massive
amounts of funding for research and development of agricultural products such as
genetically altered foods; whereas the systemic or ecological approaches which
could document any environmental implications for the use of bio-engineered
products are left with insufficient resources. As a result, potential dangers to the
human and social systems are inadequately understood as evidenced by the
international trade controversies between American exporters of genetically
engineered products and prospective customers in the European Common Market.

However, it does seem that elements of the science community are moving
to a new view of nature and science. It is a view that is more consistent with the
characteristics of Eastern thought that regards humans as integral components of the
Earth system, a view consistent with a systems approach in science. The science
education community and the current projects of renewal have not adequately
accommodated this point of view concerning the changing nature of science.
However, they need to. Future efforts at developing national curricula and an
international perspective of science literacy, must give equal attention to the
methods and products of the system sciences.

5. CONCLUSION

Atkins, et al (1997, p. 44), in reviewing selected American curriculum programs for
the OCED study, argue that changes in science curricula are needed in part, to align
curricula with "what scientists and mathematicians" study and "how they go about
conducting their inquiries". Apart from Project 2061, however, none of the curricula
reported on in the OCED study attempted to review the current nature of science.
None attempted any projection of the needed priorities for science in the future

(apart from a continued basis for technological development) and any implications for curriculum development. We contend that such a reexamination is necessary and that the systems sciences must be adequately represented in establishing a philosophical basis for science curriculum restructure.

We have attempted to document the nature of the systems science approach and the need for its inclusion in science curricula worldwide. The reasoning processes used in system science are largely absent from science curricula. Eger (1972, 1973), however, argues for the inclusion of hermeneutics in science curricula generally. As we have indicated, Frodeman (1995) considers hermeneutics a central component of system science reasoning. Donnelly (1998), in a study contrasting the teaching methods of history and science teachers in selected schools in England and Wales, concludes that certain approaches of history teachers in involving their students in making judgments concerning the interpretations of historical information are not found in science instruction within the school populations he studied. He contends that such approaches would be relevant for employment by science teachers. He relates them to current discussions of constructivist influences on science learning. We would contend that this is related to the type of hermeneutics discussed by Frodeman (1995) as a fundamental aspect of systems thinking and therefore an appropriate and necessary component of modern science curricula. However, with Frodeman actual data as reported by the scientist is the final foundation upon which the quality and accuracy of all interpretations must be judged, not simply the statements or interpretations of some other human, a historian as is the case in history, or teacher or student as in the case of history teaching.

Thus, systems thinking as we have discussed here should be included as an equal partner to the reduction type of science currently dominating curricula worldwide. How is this to be accomplished? That is the task for further research and development. Clues, however, are available now. Researchers in Japan are developing field activities that use the systems science methods (Shimono, 1998). We have observed students in Ohio enrolled in an Earth Systems Science class tracking box turtles over yearly cycles and studying their habitat by obtaining measurements of climate, soil and biological factors. Other possible approaches are the 'explanatory stories' of science recommended for British curricula (Millar and Osborne, 1998). A systems example of such an explanatory story might be an abridged version of the development of the impact theory for dinosaur extinction written by Alvarez (1997).

We also suggest that the concept of the Earth as a system, comprising many subsystems and itself a subsystem of a larger one as embodied in the ESSC Report (1988) discussed in Chapter One, is a concept that can be the theme of science curricula worldwide (Mayer, 1995). It should replace the current disciplinary approaches and many current so-called interdisciplinary approaches to science curricula with a conceptual approach that honors the important conceptual contributions of all sciences.

AKNOWLEDGEMENTS

The research for this chapter was accomplished in part while the senior author was a Senior Fulbright Researcher attached to Shizuoka University and the Institute for Research in Education of the Ministry of Education, Science and Culture, JAPAN.

REFERENCES

Alvarez, W. (1998). *T. Rex and the crater of doom*. New York: Vintage Books.
American Association for the Advancement of Science (1989). *Science for all Americans*. Washington, D. C.: AAAS.
Atkin, J. M. (1998). The OCED study of innovations in science, mathematics and technology education. *Journal of Curriculum Studies*, 30, 6, 647-660.
Atkin, J. M., Kilpatric, J., Bianchini, J. A., Helms, J. V. and Holthuis, N. I. (1997). The changing conceptions of science, mathematics, and instruction. Chapter 3 in S. A. Raizen and E. D. Britton, (Eds.), *Bold ventures: Volume 1, Patterns among U.S. innovations in science and mathematics education*. Dordrecht: Kluwer Academic Publishers.
Black, P. and Atkin, J. M., (Eds.), (1996). *Changing the subject: Innovations in science, mathematics and technology education*. New York: Routledge.
Bracey, G. W. (1998). Tinkering with TIMSS. *Phi Delta Kappan*, September, 33-36.
Capra, F. (1991). *The Taos of physics, third edition*. Boston: Shambhala.
Donnelly, J. F. (1999). Interpreting differences: the educational aims of teachers of science and history, and their implications. *Journal of Curriculum Studies*, 31, 1, 17-41.
Duschl, R. A. (1990). *Restructuring science education: The importance of theories and their development*. New York: Teachers College Press.
Earth System Science Committee, (1988). *Earth system science: A program for global change*. Washington: National Aeronautics and Space Administration.
Eger, M. (1992). Hermeneutics and science education: An introduction. *Science and Education*, 1, 4, 337-348.
Eger, M. (1993). Hermeneutics as an approach to science: Part II. *Science and Education*, 2, 4, 303-328.
Frodeman, R. (1995). Geological reasoning: Geology as an interpretive and historical science. *Geological Society of America Bulletin*, 107, 8, 960-968.
Kelves, D. J. (1987). *The physicists: The history of a scientific community in modern America*. Cambridge, Massachusetts: Harvard University Press.
Klotz, I. M. (1993). Multi cultural perspectives in science education: One prescription for failure. *Phi Delta Kappan*, November, 266-269.
Laszlo, E. (1972). *The systems view of the world*. New York: George Braziller.
National Research Council, (1993). *Solid-Earth sciences and society*. Washington, D. C.: National Academy Press.
Lubchenco, J. (1998). Entering the century of the environment: A new social contract for science. *Science*, 279, 491-497.
Matthews, M. R. (1994). *Science teaching: The role of history and philosophy of science*. New York: Routledge.Millar, R., and Osborne, J., Eds. (1998). *Beyond 2000: Science education for the future*. London: Kings College, School of Education.
Ministry of Education, Science and Culture (Monbusho), (1997), *Monbusho*. Tokyo: Monbusho.
Morowitz, H. J. (1989). Biology of a cosmological science. In J. Baird Callicott and Roger T. Ames, (Eds.), *Nature in Asian traditions of thought*. Albany, NY: University of New York.
Nakamura, H, (1980). The idea of nature, East and West. In Mortimer J. Adler, Editor in Chief, *The great ideas today, 1980*, (pp. 234-304). Chicago: Encyclopaedia Britannica, Inc.
National Research Council, (1996). *National science education standards*. Washington, D. C.: National Academy Press
Ogawa, M. (1995). Science education in a multi science perspective. *Science Education*, 79, 5, 583-593.
Shimono, H. (1998). Systemization in teaching method of outdoor education. In Rosanne W. Fortner and Victor J. Mayer, (Eds.), *Learning about the Earth as a system: Conference proceedings of the Second International Conference on Geoscience Education* (p. 13). Columbus, OH: The Ohio State University. ED 422163.

Stanley, W. B. and Brickhouse, N. W. (1994). Multiculturalism, universalism, and science education. *Science Education*, 78, 4, 387-398.

Strahler, A. N. (1992). *Understanding science*. Buffalo, NY: Prometheus Books.

Wang, J. (1998). A content examination of the TIMSS items. *Phi Delta Kappan*, September, 36-38.

SECTION TWO: APPROPRIATE LEARNING ENVIRONMENTS

Global Science Literacy (GSL) departs most markedly from other recent challenges to the science curriculum extant in most countries, in the recommended conceptual structure with its emphasis upon the organizing concept of the Earth system as discussed in the previous section. In common with current recommendations of the National Academy of Sciences in the USA, fans of Science, Technology and Science, Monbusho in Japan and the Ministry of Education of Korea among other agencies, GSL insists upon student involvement with the learning process. However, it also insists that much of this learning be acquired while in the natural environment. When I entered professional education in the late 1960s, it was the heyday of "inquiry". Every curriculum product produced with National Science Foundation funding in the United States and most adapted for or developed in other counties, insisted upon student inquiry as the basis of learning. Only by such student involvement in the science process of discovery, could students understand the nature of science, learn and use its "processes" and maintain their interest and involvement in learning about the natural world and the "product" of science that interprets that world for them. Now science educators talk about constructivist learning environments, minds-on activities, and discovery laboratories where students can construct accurate meanings of the world for themselves. Fundamentally, the goals are the same, and GSL shares those goals.

In this section, we emphasize those learning environments that have been found to be most beneficial in courses centered on Earth Systems Education and Global Science Literacy constructs. The first is borrowed from the British project *Beyond 2000*. Chris King provides a detailed set of suggestions on how to develop and present 'explanatory stories' to GSL students. Most textbooks, certainly those I have examined in the USA, fall far short of being able to elicit student interest in science, and in fact, fail to give students an adequate understanding of the nature of science and its product. We believe students should read about science in addition to experiencing it. The 'explanatory stories' approach provides an avenue and a system for developing excellent reading materials about important science concepts and discoveries. Not all ideas fundamental to science can be learned through experiences. Thus, good reading materials are essential for an effective GSL curriculum program. Along the same lines, Lillo and Lillo, a father and son team, in Chapter Nine provides a model of how readings from ancient documents, such as those of Plato and Cicero, and modern analyses of the actions of ancient cultures can be used in GSL classes. Through such readings and their analysis, students can be helped to understand the problems that humans have caused to the Earth systems through inappropriate development activities and the subsequent effects these environmental uses have had upon history.

My colleague at Ohio State, Rosanne Fortner, and I have found the "jig-saw" form of cooperative learning to be a very effective mechanism for engaging students in learning concepts, developing strategies for background research and for sharing experiences centered on learning important concepts. In Chapter Five, she discusses her experiences and provides recommendations for conducting cooperative learning classes in secondary schools. Through the process she has developed,

V.J. Mayer (ed.), Global Science Literacy, 51–52.

students emulate the scientific research teams so common now in Earth systems research, gathering data and information from many sources including the Internet, observations of nature, experiments, the library and through interviews.

In the next chapter, Slattery, Mayer and Klemm discuss the use of the Internet in conducting Earth systems investigations in the secondary school classroom and for providing up-to-date information of Earth processes and characteristics. Many sites are now available with databases that can by used by students in conducting their own investigations on Earth processes, such as earthquake occurrence, tropical storm tracking, and space exploration. In addition, there are many sites useful to the teacher and student providing background information, illustrations and diagrams of Earth systems processes and events. The Internet has become an essential classroom resource, especially for Earth Systems and Global Science Literacy curricula.

Chapters Seven and Eight illustrate ways in which students can investigate the natural environment in their vicinity. Thompson discusses the life of Charles Darwin and his contributions to our understanding of our Earth system by taking us on tour of the area of Shropshire in which Darwin was born and lived much of his life. It is an example of how the lives of famous scientists and their early experiences with nature can be used in teaching students about their environment as well as the nature and history of science. Thus, Chapter Seven provides a model of a teaching situation that can be adapted by other teachers whose schools are located near a modern or historical scientist. In Chapter Eight, Shimono and Goto, teaching in Japan and under its restricted science school schedule and urban environments provide an illustration of a student investigation, required in the school curriculum, but carried on by the student during her summer vacation, assisted by her parents. This is an avenue for providing inquiry experiences with the natural environment encouraged in Japan, and worthy of being emulated in other countries.

The last chapter by Nam and Mayer offers a philosophy on science assessment that is not unique to GSL but embodies the spirit of GSL. We believe that assessment should be a positive learning experience--one that allows both students and teachers to evaluate progress in reaching their objectives in acquiring and imparting the science of Earth systems. Students should see assessment as an opportunity to evaluate themselves and teachers, their effectiveness. We are opposed to what has been termed "high stakes testing" because of its negative effect upon both curriculum and children. In this chapter, we also discuss the implications of assessment for achieving the seven understandings underlying Earth Systems Education. This is accomplished in the context of the Korean science education evaluation system.

Taken together, the chapters in this section provide an effective description of the learning atmosphere that we would hope would be typical of curricula designed around the principles of GSL outlined in Section One.

CHAPTER 4: THE 'EXPLANATORY STORIES' APPROACH TO A CURRICULUM FOR GLOBAL SCIENCE LITERACY

Chris King, Keele University, UK

1. INTRODUCTION

In the current debate about future revision of the science curriculum in England and Wales, a document central to the discussions is *Beyond 2000: Science education for the future* (Millar and Osborne, 1998). This argues for an education for all pupils in scientific literacy. Such an education would be based on a series of carefully chosen 'explanatory stories' that take key 'big ideas' in science and explain these in a rounded, relevant and interesting way that highlights their importance in current scientific understandings and developments. Such an approach mirrors many of the aims and objectives of Global Science Literacy (GSL) and can be developed to further these aims even more effectively.

Breadth, balance, even more relevance, and more scope for the development of GSL objectives, would be added to the *Beyond 2000* proposals by adding Earth science-related stories to the list of 'explanatory stories' provided as examples in the document. This chapter provides an exemplar of how such a story entitled, 'The dynamic Earth's crust' might be developed. The story is written in an accessible, relevant and interesting way that draws together the threads of the processes that have affected our planet's evolution and will have ramifications for the future of the Earth. It is dissected to emphasize the scope for developing GSL approaches and links with other areas of science. A variety of teaching and learning activities is suggested to illustrate the range of skills and perspectives that can be taught in this way. Such a story could provide a powerful vehicle for the development of scientific understanding in young people, beyond the narrow confines of science as it is currently taught in many classrooms and laboratories. It would illustrate the key role that science can play in the safeguarding of our planet for the future.

2. REVISING THE NATIONAL CURRICULUM FOR SCIENCE

The National Curriculum (NC) is the central focus of all government schools in England and Wales for children between the ages of 5 and 16 and has now been in place for ten years. It includes the three core subjects of mathematics, English and science, and a number of foundation subjects including geography, technology, languages, physical education, etc. There were a large number of changes to the detail of the NC in the early years, such that, in science, seven different versions of the National Curriculum for Science (NCS) appeared at different times of which three were actually implemented. Since all this change caused significant problems

V.J. Mayer (ed.), Global Science Literacy, 53–78.
© 2002 *Kluwer Academic Publishers. Printed in the Netherlands.*

to the successful implementation of the National Curriculum in schools, in 1995 a five-year moratorium to changes was called. This gave five 'change-free' years for schools, as far as NC detail was concerned and five years to prepare for the changes due in September 2000. In the event, only a limited revision was carried out in 2000, with a more comprehensive review being envisaged for 2005.

In the build up to the September 2000 changes in the National Science Curriculum, a series of debates took place. The most influential of these were:

- those generated by the Association for Science Education (ASE) that resulted in the document *Science education for the year 2000 and beyond* (ASE, 1999);
- those that formed part of a seminar series funded by the Nuffield Foundation. The document produced as a result of the Nuffield seminar discussions was called *Beyond 2000: science education for the future* (Millar and Osborne, 1998).

3. THE GLOBAL SCIENCE LITERACY PERSPECTIVE

The global science literacy approach (GSL) as described in the first section of this book, focuses on:

A conceptual rather then disciplinary organization of curricula. Instead of arranging science curricula according to units or courses organized around the contributions of each of the traditional disciplines of science, GSL recommends a unified, conceptual organization of curricula. The organizational concept is the Earth as a System and its involvement in larger systems. This concept or set of concepts, after all, is the subject of all science investigations.

A broader encompassing of the spectrum of scientific methodology beyond the reductionism characteristics of current curricula. GSL includes a significant treatment of the systems methodologies of the ecologists and Earth scientists.

An inclusion of aesthetic aspects of science and its subject, the Earth System. This inclusion recognizes the incentives of many scientists in their study of Earth and brings science close to its students and their diverse cultures. It recognizes the aesthetic feelings and reactions to Earth system phenomena that are often expressed in the arts, literature and music of their nation or culture.

A recognition of the unique qualities of science and its ability to develop procedures and languages that bridge cultural and linguistic boundaries. As a component of curricula, knowledge of the culture and procedures of science can therefore help to achieve the objectives of Global Education (a Social Studies curriculum construct) thereby assisting international and inter-cultural understanding.

Science as a way of understanding ourselves as an interacting component of the Earth system that we all share and inhabit. This concept of humanity as a part of nature is consistent with many elements of Eastern thought and contrasts with the western conception of man as created separately and therefore apart from nature.

The uses of appropriate technology to assist our understanding of Earth and to conserve Earth resources. GSL does not view science as a basis for the development of technologies for defense or economic competition; a view of science that seems implicit in most of the world pre-college science curricula today.

4.   THE DEBATE ON THE FUTURE FOR SCIENCE EDUCATION – AND ITS LINKS WITH GLOBAL SCIENCE LITERACY

4.1. 'Science education for the year 2000 and beyond' and Global Science Literacy

The ASE document *Science education for the year 2000 and beyond* (ASE, 1999) was produced as the result of a wide-ranging debate across the science educational community. A constructive contribution to the debate was provided by the Earth Science Teachers' Association (Thompson, 1996). The document, produced as a result of the debate, lists a series of aims, as follows. The first two are considered important at all ages.

☐ The development of curiosity, sensitivity, social responsibility, motivation and independence in learning, when dealing with living things, the environment and the applications of science.

☐ The development of investigative observational and manipulative skills.

The following aims are considered to show increasing emphasis as learners get older.

☐ The development of scientific terminology and an understanding of key scientific ideas.

☐ Understanding of the generation, evaluation and use of evidence in making decisions, solving problems and considering scientific, personal, social and environmental issues and ethical implications.

☐ Understanding of the tentative nature of scientific knowledge and the use of theories and models in explanation.

The document notes that the science curriculum should focus on 'the place of science in our lives' in a way that encompass a holistic view of science education and enables learners:

☐ to participate fully in a technological society as informed citizens who understand the nature of scientific ideas and activity and the basis for scientific claims, and

☐ to develop intellectually and morally through experiencing the richness and excitement of exploring the natural and physical world.

This holistic view could be provided by the GSL focus on Earth as a system that would particularly offer the richness and excitement of exploring the natural and physical world and would effectively deal with living things, the environment and the applications of science. GSL emphasis on appropriate technology would form a key aspect of the pupil's view of a technological society. Both the ASE and GSL documents emphasize that pupils should see science as a way of understanding ourselves in the scientific context of the Earth environment.

Part of the ASE document (p. 4) focuses on the selection of key scientific ideas that should be included in the science curriculum. It stresses that these should:
- enable learners to make sense of science;
- have global significance (not a list of facts);
- have personal significance to the learners;
- be supported by practical activities and have applications; and
- map progression in a core of knowledge and understanding that enables children to understand how ideas develop.

The document adds: 'A number of the key scientific ideas benefit from being developed through a storyline approach'. The 'storyline approach' is explained in more detail below. All the points listed could be addressed effectively through a GSL perspective.

4.2. 'Beyond 2000' and Global Science Literacy

The Nuffield document *Beyond 2000: science education for the future* (Millar and Osborne, 1998) concludes with ten recommendations to guide future revisions of the NCS. Some were intended for the limited revision of the NCS in 2000 and some for a more comprehensive revision in 2005. The two main Nuffield recommendations relating to 2000 have now been incorporated into the revised version of the NSC. Thus, it is likely that a number of the recommendations relating to 2005 will also be incorporated at that time. Key recommendations of the Nuffield document, that relate to Global Science Literacy (GSL) but have yet to be implemented include:

- **Recommendation 1:** The science curriculum from ages 5 to 16 should be seen primarily as a course to enhance general 'scientific literacy'.
- **Recommendation two:** At Key Stage 4 (14 - 16 year olds), the structure of the science curriculum needs to differentiate more explicitly between those elements designed to enhance 'scientific literacy' and those designed as the early stages of a specialist training in science, so that the requirement for the latter does not come to distort the former.
- **Recommendation 4:** The curriculum needs to be presented clearly and simply, and its content needs to be seen to follow from the statement of aims (above). Scientific knowledge can best be presented in the curriculum as a number of key 'explanatory stories'. In addition, the curriculum should introduce young people to a number of important ideas about science.
- **Recommendation six:** The science curriculum should provide young people with an understanding of some key ideas-about-science, that is, ideas about the ways in which reliable knowledge has been, and is being, obtained.
- **Recommendation 7:** The science curriculum should encourage the use of a wide variety of teaching methods and approaches. There should be variation in the pace at which new ideas are introduced. In particular, case studies of historical and current issues should be used to consolidate understanding of the 'explanatory stories' and of key ideas-about-science, and to make it easier for teachers to match work to the needs and interests of learners.

In the light of these recommendations and those of the ASE document (ASE, 1999), it is likely that an NCS for the 'scientific literacy' of all children will be considered that may be based on some key 'explanatory stories', which include case studies of historical and current issues.

5. THE 'EXPLANATORY STORIES' APPROACH

The document *Beyond 2000* (Millar and Osborne, 1998) recommends that the science curriculum be presented, at least in part, through a series of 'explanatory stories'. The authors (p. 13) describe these as accounts that have broad features which interest and engage pupils and are able to communicate ideas in a way that makes them coherent, memorable and meaningful. The 'explanatory stories' are not fiction, but use the narrative form to present the ideas as a rounded whole. The stories (p. 13–14):

- emphasize that understanding is not of single propositions or concepts, but of inter-related sets of ideas that provide a framework for understanding;
- help to ensure that the central ideas of the curriculum are not obscured by the weight of detail so that both teachers and pupils can see clearly where the ideas are leading;
- portray the sort of understanding that one would wish young people to develop through studying the science curriculum.

The authors developed the 'explanatory stories' approach recommended in the 'Beyond 2000' document over several years. An initial impetus came from a report published by Ogborn, Brosnan and Hann (1992) that was based on a paper produced by Ogborn in 1991 (Ogborn, 1991). In these publications, the authors explored the use of explanation in science and used the term 'history' or 'explanatory history'. A 'history' is the account of a scientific situation (or 'world') from which an explanation for a scientific phenomenon is derived. This idea was built upon by Arnold and Millar (1996) in a paper containing the phrase 'Learning the scientific "story"' in the title. The authors (p. 250) state that 'Our use of the term (story) is intended to convey the complex and interrelated set of ideas which constitutes the accepted scientific explanatory framework for a particular domain of science education' before going on to discuss the 'story' of elementary thermodynamics.

This approach was further discussed by Millar (1996, p. 13) in arguing that there are a number of 'powerful models' at the heart of science that provide explanations for natural phenomena. He described these explanations as a 'story' or 'mental model' that provides a means of thinking about what is going on. He went on to 'nominate' a range of suitable models, including: the atomic model of matter; models of the solar system; a model of radiation transmission; 'field' models (gravity, magnetism, electricity); the germ theory of disease; the gene model of inheritance; Darwin's evolutionary theory; and models of the evolution of the Earth's surface (rock formation, plate tectonics). Millar felt that these models would address a number of key ideas, including those of size, scale, distance, time, cycling of materials and important applications. However, no written example of a 'story' was provided.

Meanwhile Osborne (personal communication, October 2000) has prepared an account of his understanding of the term 'explanatory stories' by reference to the literature of popular science including such authors as Carey (1995), Angier (1995) and Gould (1980). He argues that an effective narrative will begin in a context that is relevant and accessible to the hearer and imbues the story with awe and wonder. Then normal narrative devices are used to give the story interest and impact, keeping the use of scientific jargon to a minimum. He quotes research indicating that children, who had been exposed to such stories, remembered the content better than those who had been taught using ordinary scientific texts. He concludes that 'approaches that give pre-eminence to the affective components of wonder, joy and fascination, such as those found in popular science, are worthy of serious examination, and are better suited to the needs of some, if not the majority'.

The development of the idea of 'explanatory stories' should not be confused with discussions about the use of the term 'stories' in the context of historical stories of scientists and their discoveries. These discussions include those of Martin and Brouwer (1993) who suggested that through stories from the history of science 'we and our students can share in some of the doubts, struggles and rewards of scientists' (p. 458); Solomon (1993), who emphasized the value of stories from the history of science in providing an understanding of science as part of a social enterprise, and Kreiger (1992) who argued that physicists employ current 'explanatory stories' that are similar in many ways to the stories from the history of science and of biblical stories. Millar and Osborne would regard such historical stories as 'stories about the story' (personal communication, October 2000), where 'story' refers to their term 'explanatory story'. The use of narrative and the importance of stories has also been discussed in relation to other subject areas, but with a different meaning as well. In the context of geography (see McPartland, 1998) and history (see Arthur and Phillips, 2000), the 'stories' being discussed are either historical accounts, as above, or are primary sources; stories that were written at the time.

Thus the 'explanatory stories' approach of Millar and Osborne (1998) is based on Millar's 'powerful models' that see science as the development of a number of big ideas. For each big idea, an 'explanatory story' is prepared that encapsulates how the big idea developed to our present scientific understanding of it, and emphasizes the key components of the big idea and their importance for humanity today. The two exemplar 'explanatory stories' that appear in '*Beyond 2000*' (on 'the particle model of chemical reactions', p.14, and on 'the Earth and beyond', p. 16) have therefore been specially written, are relatively short (a few hundred words) and signal key concepts involved in the 'powerful model' being exemplified. They begin in an accessible way, are written clearly and simply, and attempt to explain the scientific understanding behind each of the concepts involved.

It is envisaged that a teacher would use the story as the focus for that part of the curriculum. The teacher would seek to develop an understanding in pupils of the whole story and its interlinking concepts through a number of approaches, including the use of practical work, and an emphasis on the scientific background to the story and the importance it has in the lives of people today. By selecting just a few pivotal 'stories', the overall content of the science curriculum would be reduced, thus enabling a closer focus on the development of the big ideas and their importance.

Following publication of the *'Beyond 2000'* document, the use of 'explanatory stories' in the science curriculum has been further developed in the preparation of an AS-level examination syllabus (Advanced Subsidiary GCE syllabus, aimed at 16–17 year old students) on *Science for public understanding* (AQA, 1999). This presents the content of the section entitled 'Scientific explanations' as a series of twelve explanatory stories: the particle model of chemical reactions; model of the atom; radioactivity; the radiation model of action at a distance; the field model of action at a distance; the scale, origin and future of the universe; energy, its transfer, conservation and dissipation; cells as the base units of living things; the germ theory of disease; the gene model of inheritance; the theory of evolution by natural selection; and the interdependence of species.

6. A REVISION OF THE NATIONAL SCIENCE CURRICULUM BASED ON THE 'EXPLANATORY STORY' APPROACH DELIVERED THROUGH A GLOBAL SCIENCE LITERACY CONTEXT

In the debate on future revision of the science curriculum, the challenge is to develop the curriculum in a Global Scientific Literacy context in a way that does not alienate teachers of the current curriculum, a major challenge indeed. The challenge will be addressed below by:

i)   taking the series of explanatory stories that has been suggested and attempting to recast them in a holistic GSL way.

ii)   suggesting how the GSL dimensions of one of the two 'explanatory stories' ('the Earth and beyond') published in *Beyond 2000* can be developed.

iii)   considering in detail one of the two Earth science-related stories that, it is recommended, should be added to the list of stories published in *Beyond 2000*. Part of the detailed consideration of the story shows how, having come into the story from a GSL perspective, teachers might develop its GSL focus, so coming out of the story in a GSL related way. A variety of different teaching approaches is suggested that enhance the relevance and interest of different elements of the story for both teachers and pupils.

6.1. Developing a holistic approach for the explanatory stories given in 'Beyond 2000'

The Nuffield report *Beyond 2000* (Millar and Osborne, 1998) suggests several 'quintessential stories' that merit a place in the curriculum, namely:

a.   the human body as a set of inter-related organ systems;

b.   cells as basic building blocks;

c.   adaptation of organisms;

d.   life processes in plants;
e.   mechanisms for passing characteristics from one generation to the next;
f.   evolution by natural selection;
g.   all matter being made of tiny particles;
h.   chemical reactions as particle rearrangements;
i.   different kinds of bonding;
j.   the Earth's movement relative to the Sun;
k.   the structure of the Solar System;
l.   the formation and evolution of the Earth;
m.   the structure and evolution of the universe;
n.   forces acting over long distances;
o.   the causes of motion and its control;
p.   the causes and direction of change, and
q.   radiation, light and their interaction with matter.

In order to provide breadth and balance, the following Earth science-related stories should be added:

r.   *the interrelated processes that affect the Earth; and*
s.   *global solid Earth processes explained through plate tectonics.*

(The letters a. to s. have been allocated to the stories for use in Table 1 below.)

Table 1. Holistic GSL web topics and their links

Holistic GSL web topic	Links to other holistic GSL web topics	Links to 'Beyond 2000' explanatory story suggestions
A. the land, its shape and character;	B, C, D, E, F, G, I	l, r, s
B. the air and water around;	A, C, D, E, F, G, H, I, K	l, r, s
C. their movement;	A, B, C, G, I, K	r, s, n, o, p
D. the reactions that go on within and between them;	A, B, D, F, G, H, I	l, r, g, h, i
E. natural resources and useful chemical reactions;	A, B, D, F, G, H, I	r, g, h, i
F. nearby vegetation and animal life;	A, B, D, E, G, H, I, K	b, c, d, e, f, q
G. the environment, its use and protection;	A, B, C, D, E, F, H, I, J, K	b, c, d e, f, g, h, i, p, q r, s
H. myself;	B, D, E, F, G	a, b, c, e, f
I. sources of energy and energy transmission;	A, B, C, E, G, J, K	g, h, l, m, n, r
J. harnessing energy;	A, B, C, D, E, G, I, K	h, n, o, p, q, r
K. the sun and the universe	B, C, F, I, J	d, j, k, l, m, n, o, p, q

The science of our Earth involves a complex web of interacting processes that needs to be perceived by individuals if their influence is to be positive rather than negative. If pupils can put themselves at the center of this web of processes, then the processes are given personal relevance. Probably the best way of approaching science in this way is to look out of the window, or even better, to sit on a hill or a cliff top, and to ask a series of questions, following them to their natural conclusions and often back again as well. Fruitful scientific questions in this context would relate to:

- the land, its shape and character;
- the air and water around;
- their movement;
- the reactions that go on within and between them;
- natural resources and useful chemical reactions;
- nearby vegetation and animal life;
- the environment, its use and protection;
- myself and my own role in these processes;
- sources of energy and energy transmission;
- harnessing energy; and
- the sun and the universe

This thread of topics, shown in column 1 of Table 1 could be woven together into a web through the links shown in column 2. The topics would link in turn to the *Beyond 2000* story suggestions as shown in column 3.

Trend, in Chapter Fifteen of this book, also shows how the thread of deep time is woven through these topics. There are many other ways in which topics for scientific study could be identified, interwoven and linked, but this demonstrates a GSL-related method that could be used effectively. The method can be further illustrated by taking one of the fully developed 'explanatory stories' published in '*Beyond 2000*' and dissecting it to identify its GSL potential.

6.2. The 'The Earth and beyond' story from 'Beyond 2000' dissected

The document, *Beyond 2000* (Millar and Osborne, 1988) provides two explanatory stories in detail, the second of these is 'The Earth and beyond' (p. 16). This is given below, with a commentary on each section from the GSL perspective.

Passage from the story
From our point of view of the Earth, it seems that we are living on a flat stationary surface. However, imagine moving to a point in Space, well away from the Earth. Then we would see that it is roughly a sphere which is moving in two ways. First, the Earth is spinning on an axis through its North and South Poles; this means that different parts of the Earth's surface point towards the Sun at different times, resulting in day and night. Second, it is also moving, roughly in a circle, round the Sun, taking one year to make a complete orbit. The Earth is kept in its orbit by the gravitational force between the two masses of the Sun and the Earth. Because the axis around which the Earth spins is tilted at an angle to the plane of its orbit, the relative lengths of day and night are different for the northern and southern hemispheres and, moreover, change as the Earth moves round its orbit. This is what causes seasons.

Links to global science literacy

> The Earth is part of the solar system and the interactions between the Earth and the rest of the solar system, particularly the Sun, result in day and night and the seasons with all the effects that these have on Earth and biological systems. Most biological, chemical and physical cycles are driven or are affected by the energy from the Sun and by changes cause by day/night and the seasons.

> Mentally, it is impossible to view the Earth and the Sun from afar without gaining an aesthetic appreciation of both.

> When we 'see' the Earth from afar, this puts political and cultural differences in perspective and encourages 'whole Earth' views.

> By 'looking' at the Earth from afar, we gain an impression of how the Earth fits into the solar system and how Earth systems are linked to the solar system.

> Through satellite technology, we can actually 'see' the Earth from afar today. This allows the collection of scientific data in many different ways and on a far larger scale than ever before, thus allowing a greater scientific understanding of the Earth system and its interacting processes. This therefore demonstrates the effective application of high technology in the study of Earth processes.

Passage from the story

In both our spinning and our orbital motion, we keep going at a steady speed, unlike things here on Earth, because there is no friction to slow us down. We are not the only planet going round the Sun; there are others. Three of them (Mars, Venus and Mercury) are close to the Sun like us. There are two really big ones (Jupiter and Saturn), very different from us and much further away. Finally, there are the outer ones which are very much further away and really cold. Several of the planets, including the Earth, have moons that orbit around them.

Links to global science literacy

> Understanding of how the Earth interacts with the solar system is deepened through considering how the planets differ in relation to their positions in the system.

> Planetary differences can be identified on a number of levels. Experience of Earth processes can be used in the prediction of processes on other planets.

> Consideration of the planets gives an aesthetic dimension to our lives that has inspired many artists and musicians.

> Probes used to explore the solar system exemplify the use of high technology and high levels of scientific/technological funding for academic rather than for commercial reasons.

> The success of many interplanetary probes highlights the effectiveness of scientific collaboration and the application of scientific principles to very complex problems.

Passage from the story

Of the planets, the only one with life on it (so far as we know) is the Earth. It is possible that there is life on Mars and one of the moons of Jupiter, but we don't know. If we did find life there as well, it would make the possibility of other life elsewhere in the Universe much more likely.

Links to global science literacy

➢ Consideration of possible life elsewhere in the Universe puts life on Earth, and our own existence, into perspective.

➢ Life is seen as the result of many interacting processes, many of which derive from solar processes.

➢ Possible life elsewhere in the Universe has inspired authors, artists, musicians and other people of all ages and cultures. Such considerations impinge on our own lives.

Passage from the story

Our planet is really quite unusual. While most of the Universe consists of hydrogen and helium, we live on a tiny rocky planet made out of elements which together make up less than 2% of all the matter in the Universe. Moreover, we are just sufficiently far from the Sun for water to be liquid on the majority of the surface. This has enabled life to begin. We are also big enough for there to be sufficient gravity to keep our atmosphere, unlike Mercury or the Moon.

Links to global science literacy

➢ The fact that our planet is unusual in the Universe and may even be unique gives pause for thought to all, including scientists.

➢ The unusual nature of the planet clearly shows how physical, chemical, biological, and Earth processes interact in a way that is unusual in the Universe.

➢ This also gives a perception of the fragile nature of some Earth processes and the need for sustainability.

Passage from the story

Surprisingly, the Sun is a star – a fairly ordinary, middle-aged star halfway through its lifetime and a wonderful example of a balanced nuclear fusion reaction. How do we know? Well firstly, this is the only mechanism that could possibly produce so much energy and, secondly, theoretical models based on this idea predict the behavior of the Sun quite accurately. The Sun looks bigger than all the other stars because it is much nearer. The Sun itself is just one star in a cluster of a hundred thousand million stars that we call a galaxy. You can see the cluster edge-on in the night sky as a band of stars called the 'Milky Way'. There are hundreds of millions of galaxies and these are found in clusters as well. Distances to the stars are enormous – the nearest one would take four years to reach traveling at the speed of light, and the furthest known one is 12 billion years away. So, our home, the Earth, is really just a tiny speck in an enormous Universe.

Links to global science literacy

➢ The solar system is seen as a tiny component of a galaxy system, which itself is a tiny component of the Universe.

> Our own existence and the existence of life on Earth on a 'tiny speck in the Universe' are thought provoking.
> Modeling of processes is a key activity of scientists, both through physical and computer models. Our current understanding of the Sun demonstrates the success of this approach.
> The balance that maintains the Sun can be mirrored by the balance seen in many Earth processes. Disturbing such balance can have fatal consequences.
> The awe and wonder of the contemplation of space contributes greatly to aesthetic and scientific perspectives.

6.3. Development of an Earth science-related 'explanatory story'

As noted above, in order to give breadth and balance to the list of stories recommended in *Beyond 2000*, and to provide scientific insights into the processes that affect the surface of our planet today and have caused the planet to evolve into its present state, additional Earth science-related stories need to be added to the list. One would focus on Earth processes and, in particular, on crustal processes.

The story of the evolution of the Earth's crust could be told in a number of different ways to provide different emphases. The method used in 'The dynamic Earth's crust' story described below begins in a local 'concrete' way (through the view from the beach) before developing more large scale, abstract ideas that eventually have global contexts. This approach develops the GSL potential of the story and has the characteristic of an effective narrative that begins in an interesting and accessible way and uses this as a foundation on which the story is built. A different approach could have been taken, which would see pupils themselves as part of the rock cycle, in that the elements cycling through their bodies (notably carbon, hydrogen, oxygen, sulfur and phosphorus) have, in the past, formed part of the Earth's lithosphere, hydrosphere, atmosphere and biosphere (and will again in the future) [this approach suggested by David Thompson]. A third alternative would have been to use an approach much more closely based on historical case study. Other alternatives might focus on the production and use of natural resources or on the interrelationships between the cycles that affect environments on Earth.

'The dynamic Earth's crust' explanatory story that has been prepared is given in full below, but is then dissected to show how it might be taught effectively and where links to GSL contexts are to be found. It is dissected according to the following headings:

Passage from the story - section of the story being considered;

Suitable laboratory/classroom or field activities - includes a range of teaching and learning activities that can add variety and 'bring the subject to life' for pupils;

Links across science - shows how this part of the story links to other parts of the science curriculum;

Links to global science literacy - indicates how the story can be linked outward to other key areas in a GSL context;

The Dynamic Earth's Crust - an Earth science-related explanatory story

Imagine sitting on a beach in front of a cliff of sandstone and wondering how the sandstone formed. The sand grains of the beach are the same as the sand grains in the sandstone. The sand is made by the water on the beach grinding rocks down into small fragments. Where land is sinking, in river, beach or sea areas, sand and mud can be laid down layer by layer for a long time, perhaps hundreds or thousands of years. The layers are usually flat, horizontal and widespread with younger layers laid on top of older ones. Eventually great thicknesses build up compressing the layers beneath. By this time, the layers above will have compressed the layers at the bottom and crystals will have grown between the grains from the waters that flowed through, cementing them together into a sedimentary rock. Sedimentary rocks form steadily and continuously in this way.

We can find out how ancient sediments were deposited by looking for the clues they contain and by studying how sediments are deposited today. Modern sands may have ripple marks or plant and animal fragments typical of the windy deserts, rivers, beaches or sea beds where they were laid down. Lime sands are deposited in tropical seas and most contain shelly fragments. Where salty waters dry out, evaporite minerals are precipitated. Organic materials are deposited in tropical swamps. Lavas and volcanic ash layers are found near active volcanoes. All these can be preserved in ancient rock sequences.

Many sedimentary rocks are no longer flat lying. Sometimes tilted and eroded rocks have flat-lying sedimentary rocks on top showing that the older rocks must have been tilted and eroded before more sediments were laid down on top and eventually became rocks. If the two sets of rocks contain fossils, the fossils are often different because of the great time gap, perhaps millions of years, between the laying down of the older sediment and the younger sediments. The tilting of rocks is part of large-scale folding that can sometimes be seen in small scale too, in cliffs and road cuttings. The power of the compressive forces that caused the folding can be judged from how much the original horizontal layers have been crumpled up and how much the fossils they contain have been compressed or stretched. Where the rocks could not bend, they often broke and slid to form faults. Such folded and faulted rocks are associated with ancient or modern mountain zones. Faults are caused either by compression or by 'pull apart' tensional forces or when rocks slide past one another. Many rocks also contain other cracks and fractures.

Crystalline rocks are often found where sedimentary rocks are folded and faulted. There are two major types. Igneous rocks formed from molten rock that, being less dense, rose up into and cut through the surrounding rock. Igneous rocks usually have randomly oriented crystals; the slower the cooling, the larger the crystals. They may have finer grained, chilled margins and the rocks they cut may have baked (metamorphosed) margins. If the molten rock reached the surface, it caused volcanic activity. Metamorphic rocks formed when the heat and pressure became so great that the existing rocks recrystalized (without melting) to produce new rocks with the same chemistry but different structure. At low levels of metamorphism, some original features, such as layering and fossils, are preserved as new metamorphic features (like the cleavage in slates) develop, but at high levels the rock is completely transformed, often with strong metamorphic layering or banding. Folding, faulting, metamorphism, intrusion and volcanic activity happen during mountain-building events. Some igneous and metamorphic rocks can be dated using the radioactive minerals they contain.

These rock-forming processes are linked in a dynamic cycle that relates to other key Earth cycles. *The water cycle produces the precipitation that plays a key role in weathering and results in the downhill flow of water and ice causing transportation and deposition. The rock cycle and water cycles together contribute to the recycling of the elements and compounds that form the Earth. For example, carbon is cycled through the biosphere, soil, oceans, limestone and fossil fuels. The main energy source for the biosphere and the water cycle is solar radiation whilst the internal part of the rock cycle is driven by the Earth's energy resulting from radioactive decay. A key part of the rock cycle is the plate tectonic cycle.*

6.4. The 'The dynamic Earth's crust' story dissected

Passage from the story
Imagine sitting on a beach in front of a cliff of sandstone and wondering how the sandstone formed. *The sand grains of the beach are the same as the sand grains in the sandstone. The sand is made by the water on the beach grinding rocks down into small fragments. Where land is sinking, in river, beach or sea areas, sand and mud can be laid down layer by layer for a long time, perhaps hundreds or thousands of years.*

Suitable laboratory/classroom or field activities
(Note: The activities listed have been coded, e.g. (aiii) so that the range of skills and perspectives that can be developed by the activities can be summarized, Table 2.)

- Sit on a beach (river or coastal) and ask questions. What clues relating to erosion, transport and deposition can be seen? What evidence have the processes left behind? How could ideas about the processes be tested further? If there are conservation issues, how could their effects be reduced? (ai)
- On a color image of a varied coastline taken from the web or CD ROM, estimate the percentage of coastline being subjected to erosion and deposition by measuring the total length of coastline and then measuring the total length of depositional coast (beaches, white or pale yellow, and mudflats, brown) and calculating a percentage. (aii)
- Shake small specimens of different rock types together in a plastic container to find out which are the most resistant to erosion, as described in ESTA, 1993. Which might form headlands or bays; which might form hills or valleys; which might be most valuable as building stones? (aiii)

■   Add teaspoonfuls of colored sands or gravels, muds and soils into a wide
    measuring cylinder full of water to demonstrate settling rates and layering. (aiv)

Links across science
♦   Relate the physics of movement to water currents.
♦   Consider the physical forces that result in abrasion.
♦   Interpret erosion, transportation and deposition in terms of the levels of energy
    in the transporting medium.

Links to global science literacy
➢   In a beach/cliff area, the processes being observed are the result of interactions
    between the atmosphere, hydrosphere and lithosphere and components of the
    biosphere may also be involved. Physical, chemical and biological processes are
    also interacting, indicating the integrated nature of many scientific studies.
➢   If pupils are sitting on a beach or anywhere else out of doors, and are trying to
    think scientifically, the aesthetic aspects of the environment are inescapable.
➢   Similar erosional/depositional processes can be observed occurring in all parts
    of the Earth so that scientific explanations are applicable in all cultural
    situations.
➢   In the beach/cliff situation, it is easy to consider the part that humans might play
    in the processes being observed, positive or negative.
➢   Similarly, appropriate ways of exploiting or conserving the environment, or of
    extracting energy, could be considered.

Passage from the story
*The layers are usually flat, horizontal and widespread with younger layers laid on
top of older ones.*

Suitable laboratory/classroom or field activities
■   At the beach, ask pupils if these principles apply to the beach sediments. Do
    they apply to the rock sequence forming the cliff? Would a rock fall obey the
    same principles? (bi)
■   Put several layers of red and white sand into a plastic box full of water (e.g. a
    Ferrero Rocher chocolate box). Show that the youngest is on top and that they
    are horizontal and continue to the edge of the container (unless they peter out
    first). Also, demonstrate that the layers were formed after the sand grains
    themselves (Law of Included Fragments - fragments are older) and that
    anything that cuts the layers (eg. use a ruler) is younger than the layers (Law of
    Cross-Cutting Relationships). See King and Kennett (1998, p. R8) for further
    related activities. (bii)

Passage from the story
*Eventually great thicknesses build up compressing the layers beneath. By this time,
the layers above will have compressed the layers at the bottom and crystals will
have grown between the grains from the waters that flowed through, cementing them*

together into a sedimentary rock. Sedimentary rocks form fairly steadily and continuously in this way.

Suitable laboratory/classroom or field activities

- Having shown that sand cannot be formed into sandstone by squeezing it with your hand, ask pupils to mix sand with different potential 'glues' (eg. water, brine, plaster of Paris, clay) and compress them in a plastic syringe that has had the pointed end sawn off. Then extrude the pellets and leave them to dry. Test them next day for the ones that are most resistant to erosion (Brannlund and Rhodes, 1995). (ci)
- Ask pupils to devise an investigation to find out the force at its base caused by a sequence of sandstones 1 km thick (by measuring the force applied by a column of sand in a measuring cylinder in the laboratory and then extrapolating the effects) (see King and York, 1996, p. E6). (cii)
- Calculate the pressure at a depth of 1 km given that the density of sand is around 2.5 tonnes m^{-3} (2.5 tonnes m^{-3} x 1000 m = 2 500 tonnes m^{-2}). (ciii)
- Graphs are available showing how sedimentary rock porosities reduce with depth (e.g. Edwards and King, 1999, Fig. 5.23). Ask pupils to explain these findings. How might this affect the ability of rocks to hold water, oil or gas? (civ)

Links across science

- Compression results from the downward pressure of the overlying material under the influence of gravity.
- Chemical changes result in the dissolving of minerals and their recrystalization as cements in pore spaces elsewhere.

Links to global science literacy

➢ The changes involved in compression and cementation involve pupils in thinking not only in terms of the short time-scales of their experiments but also in terms of the much longer time scales of geological events. They consider how physical and chemical processes might be seen in this light, thus giving a new perspective to these processes.

Passage from the story

We can find out how ancient sediments were deposited by looking for the clues they contain and by studying how sediments are deposited today. Modern sands may have ripple marks or plant and animal fragments typical of the windy deserts, rivers, beaches or sea beds where they were laid down. Lime sands are deposited in tropical seas and most contain shelly fragments. Where salty waters dry out, evaporite minerals are precipitated. Organic materials are deposited in tropical swamps. Lavas and volcanic ash layers are found near active volcanoes. All these can be preserved in ancient rock sequences.

Suitable laboratory/classroom or field activities

- Ask pupils to examine hand specimens of sedimentary structures and to generate different ideas about their origins. (di)
- Ask pupils to visit and/or study by video a modern sedimentary environment and contrast the features found there with those of a nearby sedimentary rock sequence. (dii)
- Form sedimentary structures in the laboratory (e.g. ripples, graded beds, see Kennett and King, 1998) to study the processes involved and then use the results to interpret ancient examples. (diii)
- Given a board on which a suitable palaeogeography is drawn in chalk, ask pupils to position specimens where they are likely to be found according to the fossils, compositions, textures and sedimentary structures they contain. (div)
- Visit and interpret different sedimentary rock sequences in the field. (dv)
- Use CD ROMS to carry out a virtual field visit to an area of sedimentary rocks. (dvi)

Links across science

- ◆ Modern and ancient environments (from at least 600 million years ago onwards) need to be considered in terms of their biological/ecological conditions as well as their physical and chemical processes.
- ◆ Deposits may be the results, mainly of biological, mainly of chemical or mainly of physical processes or a combination of any or all of these.
- ◆ Organic evolution is often closely linked to environmental evolution.

Links to global science literacy

- ➢ The fact that sedimentary or volcanic rock sequences in one region can be interpreted in terms of a range of different environments from different latitudes, altitudes and climates makes the world seem a smaller place in both scientific and cultural terms. It illustrates how understandings derived from one environment need to be applied to the study of other areas in order to build up a complete picture.

Passage from the story

Many sedimentary rocks are no longer flat-lying. Sometimes tilted and eroded rocks have flat-lying sedimentary rocks on top showing that the older rocks must have been tilted and eroded before more sediments were laid down on top and eventually became rocks.

Suitable laboratory/classroom or field activities

- Simulate the formation of an unconformity using a plastic box as described below. Compress the layers to fold them, remove the folded top layer and replace it with flat-lying layers. Ask pupils to estimate the time-gap an unconformity might represent in reality. (ei)

Passage from the story
If the two sets of rocks contain fossils, the fossils are often different because of the great time gap, perhaps millions of years, between the laying down of the older sediments and the younger sediments.

Suitable laboratory/classroom or field activities
◆ Simulate the use of fossils in correlation by using the plastic box used to demonstrate the stratigraphic principles, described above. Represent fossils of different ages by using remains of materials of different ages (e.g. fragments of old '78' records near the base, fragments of '45' long-playing records higher up, pieces of magnetic recording tape at a higher level with pieces of CD near the top and finally, pieces of minidisk). Discuss the implications, e.g. that the sequence is an evolutionary one, that the first appearances cannot be repeated, that worldwide correlations are possible, etc. (fi)
◆ Several paper-and-pencil fossil correlation exercises are available. For example, Edwards and King (1999, p. 202). (fii)

Links across science
◆ Fossils can be used in correlation in this way because of their biological evolution (cf. the technological evolution of music media used as an example above).
◆ Fossils provide critical clues to our own evolution and that of all the living and extinct organisms on Earth.

Links to global science literacy
➢ The links between fossils and evolution put life on Earth, and our part in life on Earth, into perspective.

Passage from the story
The tilting of rocks is part of large-scale folding that can sometimes be seen in small scale too, in cliffs and road cuttings. The power of the compressive forces that caused the folding can be judged from how much the original horizontal layers have been crumpled up and how much the fossils they contain have been compressed or stretched. Where the rocks could not bend, they often broke and slid to form faults. Such folded and faulted rocks are associated with ancient or modern mountain zones. Faults are caused either by compression, or by 'pull apart' tensional forces or when rocks slide past one another. Many rocks also contain other cracks and fractures.

Suitable laboratory/classroom or field activities
▪ Demonstrate folding and faulting by compression, using a method similar to James Hall's in 1815, with a plastic box (e.g. a Ferrero Rocher chocolate box or a component drawer) and a wooden partition. Add layers of sand separated by thin layers of flour. Move the partition forward to cause compression as described in ESTA, 1992 (moving it backwards will produce tensional faulting). (gi)

- Ask pupils to interpret folding and faulting in exposures, in hand specimens or in photographs or diagrams in terms of the forces and stress fields that caused them. (gii)
- Ask pupils to explain how the cross section of the island of Arran drawn by James Hutton, could be interpreted a) according to Hutton's ideas of uplift and b) according to Werner's ideas of materials being deposited from a deep ocean. (giii)
- Use pictures of fossils scanned into a computer and 'deformed' to ask pupils to interpret the forces that caused the 'deformation'. Measure amounts of 'deformation' to gain insight into the scale of the 'forces' involved. (giv)

Links across science
◆ Folding and faulting are the result of Earth materials behaving in plastic or brittle fashions, or a combination of these.
◆ Stress directions and whether they were of tensional, compressional or shear type can often be determined from the character of the deformation.
◆ If symmetrical objects such as fossils are distorted by stress (directed pressure), the amount of strain (distortion) can be calculated.

Links to global science literacy
➢ The fact that this particular part of the Earth's story has only been understood through the inputs of many scientists of many nationalities and cultures at many different times in the past is a good lesson in the multi-cultural aspects of science in the past and present. One of the reasons for this was the fact that mountains and folded rocks are found in many regions of the Earth and so attempts to explain them also developed in many regions. Other issues concerned the difficulty of observing and measuring large-scale features, the vast lengths of geological time involved and the slow rates of activity. Current global views are greatly aided by the constructive use of modern technology.

Passage from the story
Crystalline rocks are often found where sedimentary rocks are folded and faulted. There are two major types. Igneous rocks formed from molten rock that, being less dense, rose up into and cut through the surrounding rock. Igneous rocks usually have randomly orientated crystals; the slower the cooling, the larger the crystals. They may have finer grained, chilled margins and the rocks they cut may have baked (metamorphosed) margins. If the molten rock reached the surface, it caused volcanic activity.

Suitable laboratory/classroom or field activities
- Simulate intrusion and eruption in a beaker by melting red candle wax into the bottom of the beaker (about 5 mm depth), allowing it to solidify, covering it with 10 mm of washed sand and filling the beaker with cold water to near the top. Heat the beaker using a Bunsen burner and watch the eruption. This is described in more detail in Tuke (1991, p. 82). (hi)

- Examine the junctions between an igneous intrusion and the surrounding (country) rock to identify chilled margins (the edges of the intrusion that cooled faster than the rest, producing smaller crystals) and baked margins (the zone of thermal metamorphism in the rock adjacent to the intrusion). (hii)
- Use the worldwide web to plan a scientific visit to a volcano that erupted recently. Consider its distance, methods of transportation, suitable equipment, possible dangers and how to deal with them, what it might be like to be there and evidence that might be collected during the visit. Prepare a report on the plan or make a presentation. (hiii)

Links across science
- ◆ Igneous rocks can be interpreted in terms of the physical conditions of their cooling (quick cooling at the Earth's surface or slow cooling at depth due to the insulation of the overlying materials).
- ◆ The mineral make-up of igneous rocks reflects the chemistry of the original melts.

Links to global science literacy
- ➢ Scientific study of volcanicity takes place on a worldwide basis, wherever volcanic activity causes hazard and opportunity. Volcanicity is interpreted in a variety of different ways in different cultures, past and present.
- ➢ Modern technology provides an important aid to our understanding of volcanicity and its associated hazards.

Passage from the story
Metamorphic rocks formed when the heat and pressure became so great that the existing rocks recrystalized (without melting) to produce new rocks with the same chemistry but different structure. At low levels of metamorphism, some original features, such as layering and fossils, are preserved as new metamorphic features (like the cleavage in slates) develop, but at high levels the rock is completely transformed, often with strong metamorphic layering or banding. Folding, faulting, metamorphism, intrusion and volcanic activity happen during mountain-building events.

Suitable laboratory/classroom or field activities
- The use of deformed fossils to determine stress directions in low grade metamorphic rocks can be demonstrated by making Plasticene molds of fossils, distorting them and then making plaster casts. Pupils can be asked to determine the stress directions that caused the deformation. This activity is detailed in ESTA, 1990. (ji)
- Pupils can be asked to demonstrate field relationships of metamorphic rocks by placing specimens in their most likely positions on a drawing (map or cross section) of an intruded or otherwise metamorphosed area. (jii)
- The baking of clay in a kiln is artificial 'metamorphism' that demonstrates the principles of recrystalization in the solid state. (jiii)

Links across science

♦ Different metamorphic rocks are interpreted in terms of the temperatures and pressures that affected them and the chemical compositions of the original rock materials.

♦ During normal metamorphism, the bulk chemistry of the rock does not change, although different minerals do recrystalize to other minerals without melting.

♦ The fossil deformation that can be seen in low-grade metamorphic rocks can be interpreted in terms of the stresses that caused the deformation and the resultant strain.

Passage from the story
Some igneous and metamorphic rocks can be dated using the radioactive minerals they contain.

Suitable laboratory/classroom or field activities

▪ Radioactive decay and the half-life concept can be simulated using sweets, coins or other small items marked on one side, as explained in King and Kennett (1998). This exercise can be used to show how the decay can be used for absolute dating (giving a date in years, with known error). (ki)

▪ Use a Geiger Muller tube to show that common materials (eg. rocks, manufactured materials) are radioactive. (kii)

Links across science

♦ The radioactivity all around us is the result of the decay of radioactive isotopes, a continuous physical process of atomic breakdown.

♦ Radioactivity provides much of the Earth's internal energy and so has played a major role in the physical, chemical, biological evolution of our planet and in the development of its current global geography through plate tectonic processes.

Links to global science literacy

➤ Natural radioactivity is a worldwide phenomenon. It provides the energy that drives the internal part of the rock cycle and so without it, our planet would be physically 'dead'. It has real value in some circumstances such as the dating of rocks, but it can also be used inappropriately (eg. in the creation of atomic weapons).

Passage from the story
These rock-forming processes are linked in a dynamic cycle that relates to other key Earth cycles. The water cycle produces the precipitation that plays a key role in weathering and results in the downhill flow of water and ice causing transportation and deposition. The rock cycle and water cycles together contribute to the recycling of the elements and compounds that form the Earth. For example, carbon is cycled through the biosphere, soil, oceans, limestone and fossil fuels. The main energy

source for the biosphere and the water cycle is solar radiation whilst the internal part of the rock cycle is driven by the Earth's energy resulting from radioactive decay. A key part of the rock cycle is the plate tectonic cycle.

Suitable laboratory/classroom or field activities

- Pupils can be asked to plot data given to them on blank diagrams of the carbon cycle, as described in King and York (1995). (li)
- Ask pupils to draw diagrams showing how key Earth cycles interrelate. (lii)
- Invite pupils to consider how the interrelated Earth cycles would be affected if one key variable were changed, such as if solar radiation to the Earth were reduced as a result of a major ash-producing volcanic eruption or if an increase in greenhouse gases raised the temperature of the Earth. (liii)
- Consider the different ways in which Earth cycles produce economic sources of natural materials and energy supplies. (liv)
- Ask pupils to add themselves to any cycles they draw. (lv)

Links across science

- Studying the water cycle involves consideration of the states of materials and how they change; it involves kinetic theory.
- The carbon cycle has strong chemical and biological influences.
- All these cycles have crucial environmental/ecological impacts.
- Parts of all the cycles are driven by physical processes with different sources of energy.

Links to global science literacy

- The fact that all these cycles act globally shows that they need to be studied and understood globally; that affecting one part of a cycle can affect the whole and so can affect all parts of the globe.
- The fact that they interact with one another also shows that the cycles need to be studied as part of interacting systems; human intervention may affect not only a cycle but also other cycles as well.
- These perspectives illustrate the strength and weaknesses of such interacting global systems.
- Clearly, global systems can only be understood through global study.

6.5. The development of skills through the activities suggested

The activities suggested develop a wide range of skills and perspectives, as shown in the matrix in Table 2. Each of the activities suggested in the explanatory story above has been allocated a code used in the matrix. This illustrates that 'The dynamic Earth's crust' story has great potential to develop the skill base of students in a scientific context and to provide them with an understanding of scientific methodology and perspectives.

Table 2. Matrix of the skills and perspectives developed through the activities
suggested for 'The dynamic Earth's crust' story

Skill and perspectives	Activity code
Form a hypothesis	di, giii, liii
Design and plan an investigation	ai, aii, ci, cii, hiii
Conduct a practical investigation	bii, kii, cii
Carry out a practical simulation	ei, fi, gi, hi, ji, jiii, ki
Conduct an experimental investigation	aiii, aiv, ci, diii
Collect and record data	aii, aiii, giv, cii
Carry out a data plotting exercise	li
Carry out a data manipulation exercise	fii
Develop an interpretation from data provided	civ, fii, gii, ji
Carry out a decision-making exercise	div, hiii
Solve problems by applying results	dii, cii
Compile a report	hiii, mi
Carry out a calculation	aii, cii, ciii
Make a presentation	hiii, mi
Argue a case	di, div, jiii, lii, mi
Use information and communication technology (ICT) skills	ꞏ aii, dv, giv, hiii, mi
Use field skills	ai, bi, dii,, dv, hii
Develop spatial awareness (including 3D thinking)	ai, aii, bi, bii, cii, dii, dv, fii, gii, giv, hii, jii, liii, lv
Develop an understanding of time (including geological time)	ai, bi, dii, dv, ei, fi, fii, ki, lv
Develop understanding of the interrelationships of processes	ai, dii, dv, lii, liii, liv, lv
Develop awareness of health and safety issues	ai, bi, dii, dv, hii, hiii, lv
Develop awareness of conservation/sustainability issues	ai, li, lii, liii, lv
Consider the value and importance of natural resources	aiii, civ, dv, liv

Note: For more discussion on developing an understanding of geological time in pupils, see the work by Trend in Chapter 13 of this book.

7. CONCLUSIONS

The criticisms of Millar and Osborne in *Beyond 2000: Science education for the future* (1998) with respect to the National Curriculum for Science are likely to apply to many countries that have a national curriculum and many that do not as well. They conclude that the science curriculum:

- fails to equip young people to deal with scientific information in everyday contexts;
- fails to sustain their wonder and curiosity in science;
- can appear as a 'catalogue' of incoherent, irrelevant ideas;
- uses assessment contexts unlike those that young people are likely to use in later life;
- fails to address contemporary life, and

- lacks variety in teaching and learning.

They recommend that an effective method of addressing some of these problems is to devise a science curriculum addressing 'scientific literacy'. Similar moves towards a curriculum of 'scientific literacy' have guided the efforts of the American Association for the Advancement of Science (AAAS) and the National Academy of Sciences (NAS) in the development of curriculum guidelines for schools in the USA, as recorded in Chapter One of this book. Related discussions are taking place in many parts of the world (see *A comparative study on scientific literacy*, Takemura, 1999, for a review).

Millar and Osborne in *Beyond 2000* recommend that a curriculum for scientific literacy should present the science content as a series of 'explanatory stories' wherein key 'big ideas' of science are explained in narrative form. These 'explanatory stories' would show how the interrelated parts of the 'big idea' link together to provide a scientific understanding that is relevant to the pupils themselves and the world they inhabit. Such an approach lends itself to the aims and objectives of Global Science Literacy (GSL) with its emphasis on conceptual approaches to the science curriculum that encompass a broad spectrum of scientific methodology, bridging cultural and linguistic barriers, and helping us to understand ourselves in a scientific context. GSL emphasizes appropriate technology, conservation of environments and an aesthetic understanding of our planet, all aspects likely to be addressed through the 'explanatory story' approach.

To highlight these points, this chapter has taken the proposed list of 'explanatory stories' from *Beyond 2000* and has shown how they can be developed to address GSL perspectives. A detailed account of how this might be worked out is presented for one of the stories given in detail in *Beyond 2000*. However *Beyond 2000* fails to list stories of direct Earth science-related significance and thus does not include key scientific 'big ideas' that have affected our planet in the past and have direct bearing on planetary conditions, both locally and globally, now and in the future. This issue would be addressed by developing two Earth science-related stories, one concerning the evolution of the Earth's crust and the other focusing on the influence that the theory of plate tectonics has had on the understanding of our planet.

One of these suggestions has been taken as an example and an 'explanatory story' entitled 'the dynamic Earth's crust' has been developed. This is in effect, the 'story' of the rock cycle. The story has been dissected to highlight the links across science and to GSL. Such a story can be taught using a variety of teaching and learning approaches that, in turn, develop a range of skills and perspectives in pupils.

Such an approach, if implemented in England and Wales, would require no great additions to the existing National Curriculum for Science and so teachers should be able to adapt to it relatively easily. However, for it to be taught effectively, some form of continuing professional development (termed INSET, In-Service Education and Training, in Britain) would be necessary. Such INSET would equally be necessary if a strategy based on 'explanatory stories' were to be applied across the whole of the science curriculum.

Thus the document *Beyond 2000* provides a series of recommendations that, if built upon effectively in a Global Science Literacy context, would result in a curriculum that would not only address the problems listed above, but would also provide a fascinating and relevant science education for the young people who will become the guardians of the future of our planet.

ACKNOWLEDGEMENTS

Many thanks to David Thompson for his support and for his critical reading and many contributions to earlier drafts of this manuscript. Also to Robin Millar and Jonathan Osborne for their personal communications on the topic of 'explanatory stories'.

REFERENCES

References for the body of the chapter

Angier, N. (1995). *The beauty of the beastly: new views on the nature of life.* New York: Little and Brown.
Arnold, M. and Millar, R. (1996). Learning the scientific "story": a case study in the teaching and learning of elementary thermodynamics. *Science education,* 80, 249 – 281.
Arthur, J. and Phillips, R. (2000). *Issues in history teaching.* London: Routledge.
Assessment and Qualifications Alliance (1999). *General certificate of education, advanced subsidiary: Science for public understanding.* Manchester: AQA.
Association for Science Education. (1999). *Science education for the year 2000 and beyond.* Education in Science, No. 181.
Carey, J. (1995). *The Faber book of science.* London: Faber.
Gould, S. J. (1980). *Ever since Darwin: reflections in natural history.* London: Pelican.
Krieger, M. H. (1992). *Doing physics: how physicists take hold of the world.* Bloomington: Indiana University Press.
Martin, B. and Brouwer, W. (1993). Exploring personal science. *Science Education,* 77, 441-459.
McPartland, M. (1998). The use of narrative in geography teaching. *The Curriculum Journal,* 9, 341-355.
Millar, R. (1996). Towards a science curriculum for public understanding. *School Science Review* 77, 7-18.
Millar, R. and Osborne, J. (1998). *Beyond 2000: Science Education for the Future.* London: King's College, University of London.
Ogborn, J. (1991). *Describing Explanation.* ESPRIT II Basic Research Actions Working Group 6237: Children's and Teachers' Explanations, Technical paper number 1.
Ogborn, J., Brosnan, T. and Hann, K. (1992). *CHATTS Working paper 4. Explanation: a theoretical framework.* London: Institute of Education.
Solomon, J. (1993). *Teaching science technology and society.* Oxford: Oxford University Press.
Takemura, S. (1999). *A comparative study on scientific literacy.* Hiroshima: Hiroshima University.
Thompson, D. B. (1996). Science education for the 21st century – an ESTA response to the ASE debate. *Teaching Earth Sciences* 21, 83–88.

References for the 'dynamic Earth's crust story'

Brannlund, P. and Rhodes, A. (1995). *How the Earth works: a science teacher's guide to essential Earth science at Key Stage 3.* London: The Geological Society.
Earth Science Teachers' Association. (1990). *Hidden changes in the Earth.* Sheffield: ESTA, Geo Supplies.
Earth Science Teachers' Association. (1992). *Earth's surface features.* Sheffield: ESTA, Geo Supplies.
Earth Science Teachers' Association. (1993). *Sediment on the move.* Sheffield: ESTA, Geo Supplies.
Greene, M. T. (1982). *Geology in the nineteenth century.* London: Cornell University Press.
Hallam, A. (1989). *Great Geological Controversies.* Oxford: Oxford University Press.

King, C. and Kennett, P. (1998). *SoE3: Geological Changes - rock formation and deformation.* Sheffield: ESTA, Geo Supplies.

King, C. and York, P. (1995). *SoE1: Changes to the atmosphere.* Sheffield: ESTA, Geo Supplies.

King, C. and York, P. (1996). *SoE2: Geological changes–Earth's structure and plate tectonics.* Sheffield: ESTA, Geo Supplies.

Edwards, D. and King, C. (1999). *Geoscience: understanding geological processes.* London: Hodder and Stoughton.

Tuke, M. (1991). *Earth science: activities and demonstrations.* London: John Murray.

CHAPTER 5: COOPERATIVE LEARNING: A BASIC INSTRUCTIONAL METHODOLOGY FOR GLOBAL SCIENCE LITERACY

Rosanne W. Fortner, The Ohio State University, USA

1. INTRODUCTION

The development and implementation of GSL curricula in countries around the world could potentially have a positive impact on citizens' worldviews and understanding of other cultures. Citizens exposed to such curricula should be able to engage more effectively in a world approaching globalization through politics and commerce. Importantly for the desired outcomes of GSL, those who have internalized the curricula should see Earth's environment as one without borders. They should see the need for working together for the common goal of a sustaining and sustainable environment.

One of the most important issues in world affairs is how to establish cooperation in an atmosphere in which competition has been the norm for centuries. When the world was not competing for land area with which to expand its empires, it was competing for the wealth of the lands. When individuals were deprived of competition, as in communist regimes, the quality of their performance was prone to decline, for the measure of excellence ceased to be based on comparative quality. As advertisers compete for buyers, and as organizations compete for members, there are expressions of one entity's benefits compared to the other. When many are competing, there must be "play-offs" to see who is ultimately the best.

Education has unfortunately been no different. Science educators, politicians, the mass media and parents around the world have taken great interest in the tests comparing students' science and mathematics achievement in different countries. No matter that the educational systems in those countries are very diverse, and individual goals may be culture-bound; higher scores are interpreted by most as signaling "better" science. Should we compete for science literacy, or cooperate to achieve it?

In classrooms, we have competed against an arbitrary scale determined by a person assumed to know "the answers" to questions. If all students excelled, the questions must not have been difficult enough. A teacher who assigns a large number of high grades is assumed not to be teaching on a high enough level. Low grades are a sign of rigorous instruction, through which the ignorant are raised (not quite) to the level of their master. All are provided with the information; those who can remember it for the examination are the winners. Thus, one student is at the head of the class, because s/he competes best. How, then, is a student to be taught the value of *cooperative* learning, when the measures of excellence are the traditional *competitive* ones?

79

V.J. Mayer (ed.), Global Science Literacy, 79–92.
© 2002 *Kluwer Academic Publishers. Printed in the Netherlands.*

2. VALUES OF GROUP LEARNING/COLLABORATION FOR LEARNING
SCIENCE

Imagine how different the world would be if people were taught from the first days of life that the sharing they do on the playground would also serve them well in creating a peaceful world and an educational system fostering human development in its most positive patterns! In the essay, "All I really need to know, I learned in Kindergarten" (Fulghum, 1993), we are reminded of the importance of dealing with others in a simple, forthright manner, and of the benefits that come from this behavior. Some of Fulghum's bits of wisdom with application to world cooperation are: Share everything. Play fair. Don't hit people. Clean up your own mess. Say you're sorry when you hurt somebody. My personal favorite is "When you go out into the world, watch out for traffic, hold hands, and stick together." If we work together and use our collective wisdom, we can survive potentially harmful situations.

We can also apply some of these maxims to the learning of science. We can't all know it all, so why not share information, build each other's competencies, and grow together? If we trust each other to do our best and share our talents, we not only gain allies instead of enemies; we also build up a collective body of knowledge and experience larger than our own. We have a bigger bag of tricks when it comes to figuring out the answers to complex questions, and we have a nurturing environment that will not let us quit when there is a chance for success.

In environmental education, the emphasis is usually not on competition but on accomplishment through cooperation and collective action. We teach about the Earth as a system, with interacting components that are always affecting each other (Fortner, 1991). We cannot study individual disciplines without seeing how they are connected to other disciplines. Since 1988, the Earth Systems Education program at The Ohio State University and in many other areas has been developing curriculum materials and cooperative instructional methods for the multi-disciplinary sciences of global change and other environmental issues (Fortner, 1996). The combination of curriculum development and teacher education provides schools with high profile examples of the ways humans have affected the Earth system and how they must cooperate to restore and maintain a sustainable environment for the future.

Thus, it is natural that Global Science Literacy relies on cooperative learning as a prominent feature of instruction. Slavin (1983) defined cooperative learning as referring to "instructional methods in which students of all performance levels work together in small groups toward a group goal. The essential feature of cooperative learning is that the success of one student helps other students to be successful." While the concept of cooperation in teams is not new, research is building to justify its use in instruction through cognitive and social gains, and the methodology of cooperation is consistent with goals of science in society.

In a special issue of *Theory Into Practice (TIP)*, the College of Education at The Ohio State University has focused on "Building Community Through Cooperative Learning," and that is exactly why global science literacy depends on the collaborative process. In the issue, the co-directors of the Cooperative Learning

Center at the University of Minnesota, David W. Johnson and Roger T. Johnson (1999), discuss what is and what is not cooperative learning:

> Cooperation is working together to accomplish shared goals and cooperative learning is the instructional use of small groups so that students work together to maximize their own and each other's learning. Within cooperative learning groups, students are given two responsibilities: To learn the assigned material and make sure that all other members of their group do likewise.

Assessment of student outcomes from such lessons takes into account their non-competitive nature. Thus, individual achievement is measured by portfolios, concept maps, and other alternatives to testing, and group achievement is observable as a communicated product that synthesizes the group's activities, research, or thinking about a topic. Research in many situations has shown that cooperative learning usually produces similar or greater gains in knowledge in comparison to traditional methods (e.g., Allen and VanSickle, 1984; Humphreys et al., 1982; Okebukola, 1985; Slavin and Karweit, 1985). It is important to note that social gains are included in the assessment of cooperative learning outcomes in most of the recent research (e.g. Slavin and Fashola, 1998; Stevens and Slavin, 1995). It is those aspects that are most exciting to those who look toward schools as providing teachable moments for more than just college entrance examinations.

3. TYPES OF COOPERATIVE LEARNING

Not all groups assembled in a classroom are cooperative. As Johnson and Johnson (1999) describe: There are pseudo learning groups in which students assigned to a group have no interest in or incentive for learning together, and believe they will still be ranked as highest to lowest. As a result of grouping the sum is less than the potential of the individual members; group activity hinders the learning process. Traditional classroom learning groups accept that they must work together, but group growth is not an internal goal. Some students let others do most of the work, and the workers are frustrated and feel exploited. In such cases the conscientious students would be better achievers if they worked alone, and some students do not work at all yet get marks for completion.

In a true cooperative learning group,

> students work together to accomplish shared goals. Students seek outcomes that are beneficial to all. Students discuss material with each other, help one another understand it, and encourage each other to work hard. The result is that the group is more than a sum of its parts, and all students perform higher academically than if they worked alone" (Johnson and Johnson, 1999).

Three general types of true cooperative learning have been described by Johnson, Johnson and Holubek (1998):
- Formal (for teaching specific content);
- Informal (to insure active cognitive processing of information during a lecture or demonstration; and
- Cooperative base groups (to provide long-term support and assistance for academic progress.

Cooperative learning strategies place the responsibility for learning more squarely on the shoulders of students than do the lecture-discussion strategies that dominate college classrooms. This does not imply that the responsibility of the instructional leader is diminished. On the contrary, the design of cooperative learning experiences is more difficult than lecture preparation, as it requires a high degree of advance organization, anticipation of student response, and concurrent development of meaningful applications of the subject matter, for use in assessment.

Cooperative learning has been used in many settings for many different purposes. While the beginnings of cooperative learning probably extend back to the division of labor among pre-hominids, more recent applications in education often cite the work of Johnson and Johnson (1975/1994) for the basis of the ideas. There are many strategies for implementation of cooperative groups, and the books by Johnson and Johnson detail the components and strengths of those. Our experiences in Earth systems education have tested various cooperative learning methods and settings. The form of cooperative learning I use most often is the "jigsaw" (Johnson and Johnson, 1975/1994). This has worked well in our teacher education programs (e.g. Mayer, Fortner and Hoyt, 1995), where teachers have a stake in learning about and using the technique. Briefly, the class is divided into groups of six students or fewer, based on some characteristic that gives them something in common as a basis for discussion. These are the Base Groups. To reinforce their common characteristics, they select a group name. This identity will shape their activities within the jigsaw, as they will be encouraged to apply new learning in the context of their Base Group's needs.

The new learning comes with the activities of Expert Groups. Members of the Base Groups, after preliminary discussions that establish a common background, are divided into working teams that include one or two members of each Base Group. This new group becomes an Expert Group on one component of the subject matter. Activities of Expert Groups are structured by the teacher to assure that certain objectives are met and each group member contributes to the learning experience. Upon completion of those activities, the Experts reassemble into their original Base groups to teach their peers what they have learned. Each student's responsibility to both groups is clear, a combination of learning and peer teaching for a common specified goal.

Base Groups include students with common characteristic as a basis for applying information (major, career goal, etc).

1 1 1	2 2 2	3 3 3	4 4 4	5 5 5	6 6 6
1 1 1	2 2 2	3 3 3	4 4 4	5 5 5	6 6 6

Expert Groups contain at least one member of each Base Group.

1 2 3	1 2 3	1 2 3	1 2 3	1 2 3	1 2 3
4 5 6	4 5 6	4 5 6	4 5 6	4 5 6	4 5 6

Experts return to Base Groups to share information for common needs.

| 1 1 1 | 2 2 2 | 3 3 3 | 4 4 4 | 5 5 5 | 6 6 6 |
| 1 1 1 | 2 2 2 | 3 3 3 | 4 44 | 5 5 5 | 6 6 6 |

Figure 1. Organization of students in a jigsaw cooperative learning experience. Each number represents one student in a class of 36

4. EXAMPLES OF GROUP PROCESS WITH EFFECTIVE RESULTS

In the work of Earth Systems Education, we have used cooperative learning in a number of ways. Described here are examples of how jigsaw techniques have been used at three levels of learning to help develop global science literacy.

4.1. Undergraduates learning about their professional literature

While undergraduate students are learning the basics of their sciences and how those information components are combined into lessons, I have them examine the literature of the field with their own professional needs in mind (Fortner, 1999). The genres of science literature are many, but as college students, I invite their attention to the primary science reporting, secondary features based on those writings, television treatment of science, and popular writings for practitioners (educators, in this case). The jigsaw process is most effective for such an activity. The purposes of this group process are several:

- Expanding the amount of science materials to be introduced in a short time.
- Providing people with perspectives similar to their own (rather than from the instructor's viewpoint).
- Introducing new professionals to literature they may want to have in their own professional libraries.

The procedure followed is summarized below. This information has been previously reported in the *Journal of College Science Teaching* (Fortner, 1999) and contains Johnson and Johnson's (1999) structural steps for formal cooperative learning: Pre-instructional decisions, Explanation; Monitoring; and Assessment.

> An initial brainstorming session with the full class is used to focus attention on how [the profession] uses its literature. Discussion includes the primary literature of science, interpretive literature for various target groups who need the science information (such as teachers, recreationists, researchers, etc), information produced by special interest groups, environmental news as part of other news coverage, and the like. We attempt to cover as many sources and recipients of environmental information as the students can envision.

> In the meantime I collect from colleagues, home, and students at least 5-6 samples of publications that represent each category of literature. Each category will be used in a separate jigsaw, so the categories will need to include as many different publications as there are Expert Groups. For example, a category of "Science Background" might include publications best used by an informed audience. In other words, to get the most from this category of science literature, a reader would have to know some science first. Based on availability such a set of literature could encompass *Science, Bioscience,*

Nature, Journal of the American Medical Association, Climate Change, American Scientist, or perhaps such secondary but science-rich works as *Scientific American, Earth,* and *Discover.*

In the Expert phase of the jigsaw, each group receives copies of different issues of ONE publication in the category. They also have a preliminary list of suggestions for their out-of-class review. This includes some general items ... plus specific items related to the instructional objective. Students review their individual copy of the publication for one week. In a class of 30, there are six issues of five different publications being examined over the week.

On the day of the synthesis and debriefing on the category of literature, Experts meet in their groups for about 30 minutes to synthesize their findings about the publications. It may be useful to have a new set of focus questions available at this time so the synthesis is expedited. It is important that each Expert Group agrees internally about the important points to be shared regarding their publication. Experts are essentially preparing to teach their Base Groups, and individual Expert members will be alone in their Base Group to do this teaching. Personal responsibility for learning and for teaching quickly become apparent to the students.

After the synthesis period, I use the rest of the two-hour class block for Base Groups. Students reassemble so that each student with a sample of the Expert publication comes together with 4-5 others who have become experts on different titles. Again there is a common set of debriefing questions so the students get a comparison of basic characteristics of the publications. Some make personal charts to remind them of this information later when the other Experts are not available. Ample time is left for the students to discuss the styles of the journals, scope of subject matter, reader reactions, what a reader is assumed to know, and such characteristics that would help a new professional select one for a subscription.

Several important processes and outcomes are inherent in this exercise, and it is critical that students realize they are being evaluated on their performance within each. This must not be treated as a testable course component for which right answers are expected, but grades do have to be assigned for the course. Because students expect testing and right answers, many are confused and uncomfortable in a cooperative learning situation. For the first few experiences with the process, I try to increase comfort and decrease confusion by giving more direction to the phases of the jigsaw, as noted above. In later applications of the process, students know what to expect and are convinced that the system is fair and valuable, so fewer guidelines are needed for keeping the experience moving.

Time Utilization	Needs continual reminders to get on task and on topic. Often late meeting requirements.	Usually on task and topic, but needs occasional reminders. Occasionally late.	Always on task, on topic, and on time with appropriate work.
Participation in group effort	Does not add equitable amount of work to the effort	Adds an equitable amount of input, but may not meet requirements of tasks	Adds an equitable amount of input and meets or exceeds requirements of tasks
Accuracy of Information	Information shared is inaccurate or lacking in content basis	Information is usually correct but may not be relevant to group needs	Information is factually correct and pertinent to the group's needs

Clarity of information	Student is not organized, and information shared is not clear	Student work is well planned but not always clear	Work is well planned and clearly explained. In command of the situation and its needs

Figure 2. Sample rubric for evaluation of individual student input to group process

For the group processes themselves, I use a simple rubric (Figure 2) that shows the students the possible levels of behavior and how those are valued in a grading system. I try to apply the rubric twice during the synthesis and debriefing session, once while Expert Groups are assembled, and once with Base Groups. Groups frequently become adept and honest with internally rating their group members as well, and prefer to do the evaluation themselves in later applications of the process. On other occasions, interest is so high that individual evaluations do not appear necessary.

To culminate the experience with a particular category of literature, individual students are asked to construct a portfolio element in which they select one of the titles to which they would likely subscribe, plus an alternate that could be justified under some conditions. The students write a one-page rationale for their choices, using parallel criteria of professional need and publication characteristics. The portfolio elements are evaluated using a separate rubric.

The reading assignments are received well by the students because the individuals are learning about their own professions before they have to practice them, and because they can't afford to read all of the journals themselves just to decide which one suits their needs best. The sharing of perspectives among peers becomes an important outcome of the classroom activity. Individuals' readings are combined to discuss what is understood and what remains to be asked. The same can be accomplished with the content of the science as well as the content array within journals.

4.2. Teacher education for self-assessment of content and needs

In some of the teacher education programs we have developed with support of the National Science Foundation and Sea Grant, we have required teachers to study original writings first to assess their own informational level and needs before being invited to do classroom activities or to talk with a scientist who wrote the materials (Mayer, Fortner and Hoyt, 1995). In the NSF workshops of the Program for Leadership in Earth Systems Education, we used the following method to update teachers' understanding of current news in the field. Global change and the use of modern technologies were the focus of the selected science, and topics included the following, involving three scientists/topics per summer: tektites, trees' response to climate change, ice core data, global change policy, hydrothermal vents, and El Niño effects on the Galapagos.

The teachers first read articles by the scientists and discussed them in mixed-grade groups so that the more knowledgeable teachers could assist others in understanding the material. Once the teachers have studied the original writings or read the chapters of a reference text, they are presented with some questions that probe the depths of their understanding of the material. These are not questions with exact answers within the readings, but challenges to their understanding, requiring analysis and synthesis of information. By discussing the information within groups that have a similar background or similar needs for teaching, their understanding of the material is constructed within their own experiential context. The subject is discussed on that level for maximum understanding and a view toward how they can use the information. Frequently leaders will arise within the groups to direct the flow of discussion, but nevertheless the responsibility for learning is shared based on a common experience.

Scientists later came into these groups of ten or fewer teachers to discuss the subject and its global importance with them. Teachers then met in grade level groups (e.g. all elementary together) to talk over the information from the scientists and develop questions on things they still did not understand or aspects they wanted to pursue further. Scientists met with these grade level groups and clarified information again, this time specifically on the level that a teacher might use it. In this way, the teachers came to see the scientists as real people who were excited about their work and who also cared about how that work got into the curriculum. The scientists in turn gained great respect for the teaching situations related by the teachers, and for the level of understanding and interest teachers had in the Earth systems subject matter. The scientists did not lecture but instead spent three days cooperating in the learning experiences with the teachers.

4.3. Middle school science application, for combining information effectively

For most individuals, learning is assisted through cooperation. More is accomplished because tasks are shared, the large task of learning is broken into individual chunks. As teachers, with large amounts of material to "cover," grouping lessons not only helps students see the relationships between experiences but also assists in getting a wider array of information put before the students for real use.

A project at The Ohio State University helps students combine real environmental data from different sources to bring to discussions of big questions about Earth systems. Working cooperatively, students seek the data that will help them understand the complexity of the problem and what it would take to find a reasonable answer. A "tool kit" of resources on the Great Lakes Solution Seeker CD-ROM gives a starting place for investigations, and hard-copy materials offer guidelines. An example is given here for the question, What factors are at work to erode coastlines in Lake Erie? (Fortner, 1995). The datasets would be used as follows; each used by a different Expert group and then combined for considering the question.

- Aerial photos can show where erosion is occurring and how much. This gets at the sub-questions of how lithosphere conditions affect the problem, and how big the problem is. Many areas along the Great Lakes are sites of dramatic erosion because of their composition, slope and other physical factors. Human developments exacerbate natural processes by changing runoff amounts and patterns, as well as removing protective vegetation. The data for studying a classic site on Lake Erie are available from state records, from time-series aerial photos, and from an Internet live cam.

Using the aerial photos, students draw the older shoreline and a more recent one to get the surface area of land lost. To find the volume of land lost, a topographic map is examined to get the height of the bluff. In this example the area east of the groins lost land at the rate of 22,737 cu ft/year, while west of the groins (up current) the rate was 8421 cu ft/year.

- What is the pattern of storms on Lake Erie? This sub question leads to consideration of wind speeds, fetch sizes, wave heights and storm surge magnitudes. All can be investigated using the Great Lakes Forecasting System, an Internet site [http://superior.eng.ohio-state.edu] that provides near real-time data and archives of physical factors that might affect erosion rates at different sites along Lake Erie.

- How are precipitation and water levels related? Do either have the potential to contribute to erosion? These questions are investigated through databases maintained by the National Climate Data Center and the U.S. Army Corps of Engineers. Comparison of the data requires that students use graphing skills, select data from huge amounts available, and determine the most effective methods for arraying information to search for trends and relationships.

This example is from ES-EAGLS instructional materials produced by Ohio Sea Grant and available from the chapter author. Activities in this set of five booklets can be grouped for cooperative learning to answer many other kinds of important Earth systems questions. The books are used throughout the region and in numerous other areas of the US and other countries. The concepts they address are common to the middle grade science curriculum, and the local (Great Lakes) examples increase their relevance for students. For GSL they serve as examples of an investigative, interdisciplinary approach to science learning, and as methods of grouping classroom activities for varied goals through cooperative learning.

4.4. High school course combining jigsaws and other cooperative learning

The Biological and Earth Systems Science (BESS) course is a two-year laboratory science curriculum in the Worthington (OH) City Schools (Fortner, et al, 1992). Students typically take the course in grades nine and ten. The program is a diverse one designed to foster growth not only in science learning but also in use of science knowledge and establishment of social skills in teamwork. Much of the work of the course is carried out in teams, not always as one of the cooperative learning models but at least requiring participation and interdependence for accomplishment of tasks. The staffs of the schools have worked hard to structure the cooperative activities in ways that eliminate the incentive to compete. Ability levels within the heterogeneous classrooms could easily become an issue if "Traditional classroom groups" were the operational mode. Instead, the twelve-year-old program is selected over traditional biology by about 90% of students. Its students have maintained their excellent ratings on state measures of achievement even though competitive testing is uncommon along the road to those measures.

One activity developed for the *Activities for the Changing Earth System* (1993) is illustrative of how students accustomed to cooperative learning might use their knowledge. The "Global Climate Game" makes an excellent culminating activity for studies of global change. Teams of students move a single playing piece around a game board, making collective decisions in response to issues that will affect the amount of CO_2 in the atmosphere. For each decision, environmental points are awarded, and amounts of CO_2 increase result. With the object of the game being to have the least effect on global change, student teams soon realize that making wise choices in personal lifestyles will have the most positive effects, but they also find that not everyone in the team may agree with those choices. Since the game, as in life, reminds them that "we're all in this together," they evaluate costs and benefits of options without seeing them as simply good or bad. This is the ultimate value for cooperative learning (Fortner and Mayer, 1997), and an ultimate goal of GSL

5. RESPONSES TO COOPERATIVE LEARNING

The most important outcomes of cooperative learning experiences include some that do not typically accrue in classes:

- the real-world aspect of not having exact guidelines for production and progress;
- the expectation that groups will work together to accomplish a task, and each member's input is supported as needed so each may gain its maximum value;
- realization that the best method of learning is teaching;
- acceptance of the idea that what is gained by an individual has an intrinsic reward that may not be the same gain or reward as for another individual;
- recognition that not everything that counts can be counted, and not everything that can be counted counts.

5.1. Undergraduates

The professional journal activity was designed to introduce the types of literature that professionals should be aware of and use. Overheard discussions and requests to borrow the journals for longer periods after class was completed demonstrate that the desired interest was piqued among the group.

Student evaluation of the experiences at the end of the course indicated that the seniors were more enthusiastic about the alternative learning experiences (jigsaw sessions) than about the lectures that made up another component of the course. However, they felt they learned about the same amounts of new information from each mode of instruction. Some students remained uncomfortable with what they felt was a lack of direction in the jigsaws; they asked for some rules and for lists of information they were expected to derive. Verbal feedback in an exit discussion with the class indicated that the jigsaw and portfolios caused the students to do more work for the course, and most felt they learned the information better. For undergraduates, it is important to be clear that flexibility is intentional and completeness within the time available is expected. Utility to the individual is a primary goal, and each individual demonstrates that in the portfolio assessment.

5.2. Teacher response as cooperative learners

It is not surprising that teachers, too, feel uncomfortable the first time they are brought into a cooperative learning mode. We have all learned in the same competitive system, and we all expect that there are correct answers we are supposed to uncover. If the perceived correctness is hard to decipher, or if various levels of correctness are acceptable, this creates some dissonance. Teachers, after all, aspire and are expected to be models of high achieving, and new rules for achieving take time to accommodate.

In intensive workshops for teachers (2-6 days) it may not be possible to make participants completely comfortable with cooperative learning as the basis for a grade. Teachers in short science methods courses at our field station have mixed reactions to the use of the technique. First, they really would like to experience all of the parts of the Expert learning (i.e. try all the labs and activities themselves rather than being told about them by others). This is probably good pedagogical thinking, since the experience is the best teacher. If the things they learn from an Expert group lead them to try the labs themselves, the situation will be successful, but if they avoid labs they have not experienced then they will be missing valuable instructional opportunities. On the other hand, the basic idea in cooperative learning is to be able to trust one's peers to provide accurate and usable information. Those who object to dividing the responsibilities have missed that goal.

Second, the teachers would really like to have a lecture occasionally, and the response to infrequent opportunities for this is marked. There may not be audible sighs of relief when a lecture time is announced, but at least there is a visible level of comfort that develops from "relaxing" into the information-receiving mode. Some openly express disappointment that they do not get to hear more such lectures from someone who is well known in the field. It seems they rate my accountability lower if I give them responsibility for their learning. As the instructional leader, I must reach a balance with my approach that accepts some of those negative feelings and responds to the concern by assuring teachers they are getting "the right stuff" even if I am not outlining it on the board. I do this through evaluations, frequent discussions, and giving the teachers the opportunity through projects and short-term tasks to demonstrate their successful learning.

5.3. Teacher use of the model after a workshop

Do teachers use cooperative learning after they have experienced it in our in-service programs? Evaluation of the NSF Program for Leadership in Earth Systems Education (PLESE) indicates that teachers felt they had increased their understanding of the science topics, and the combination of approaches was effective for them. They would not have preferred simply listening to the scientists as a mode of learning (Fortner and Boyd, 1995). The teachers also responded to the following questions on a scale of 2 = Strongly Agree, 0 = No Opinion, and -2 = Strongly Disagree.

The approach was an effective way for me to learn about the science topics.	1.28
I plan to increase my use of collaborative learning as a result of this workshop.	1.07
I would like to learn science content in the future using this type of collaborative learning.	1.02
I would have understood the topics more clearly if all the available time was used by the scientist lecturing to the group.	-0.53

In addition, 74% of the responding teachers agreed or strongly agreed with the statement "I have increased my use of collaborative or group learning with my students as a result of my participation in PLESE."

5.4. High school student/parent response

Interestingly, middle school teachers and their schools report few concerns of the type described. In the BESS classes that are heterogeneously grouped, the students who have excelled in traditional systems (lectures, followed by tests) are very uncomfortable with relying on other students for part of their information. Correct answers for them are to come from the teacher, who will in turn make sure they know enough correct answers to go forward. While they may initially enjoy the social setting for cooperative learning, these traditional achievers become increasingly unnerved when some members of their group work less and provide little support for the group effort. The high achievers, unwilling to let the effort fail, take up the heavier load and carry it. As a result, their team becomes dependent and the achiever becomes angry and frustrated. Naturally, the teacher needs to watch for such situations and help correct them before mutiny occurs. In more homogeneous classrooms, fewer such issues emerge. Unfortunately, the "real world" workplaces that students will enter are likely to be heterogeneous, and unless group development skills are built and practiced, frustration will continue after the class is over.

High schools attempting cooperative learning may feel pressure from parents as they strive to meet those competitive criteria they feel will drive their student's future success. The high achievers described above may paint a classroom scenario for their parents that in no way resembles what that older generation recalls as effective settings for learning. There is therefore a need for teachers to proactively make students and parents aware of reasons for using cooperative learning, and values to be gained from it. It is also necessary to provide students with adequate "debriefing" after cooperative learning, so they are assured that they have assembled a useful set of working information for proceeding into the next areas of study. Praise for multiple working hypotheses, examples from genuine science situations, and introduction of expert groups of scientists may assist in making students and their parents more comfortable with cooperation as a means of academic success.

6. SUMMARY: IMPORTANCE OF COOPERATION/COLLABORATION FOR GSL

In science, as in most global endeavors, achievement is typically measured in comparison to how others achieve. We compete to be the best, to be first, to "win." Competition among humans has historically been one of the roots of environmental degradation. The "tragedy of the commons" is a well-known example; war is another. Any time we strive to achieve at the expense of others we may not only damage those resources on which we depend, but also cancel opportunities to work together with those for whom collective wisdom and power might bring even greater rewards. In the end, no one wins at all.

As we strive to achieve global science literacy, let us look at the examples of success from cooperative learning in science. The lessons are as those in kindergarten, but applied in grown-up and real world terms. We must demonstrate in as many ways as possible, to our students, their parents, administrators, and other educators, the benefits of cooperation as a means of achieving important goals for the environment and society.

REFERENCES

Allen, W. H. and VanSickle, R. L. (1984). Learning teams and low achievers. *Social Education*, 48, 1, 60-64.

Aronson, E. (1978). *The jigsaw classroom*. Beverly Hills, CA: Sage.

Fortner, R. W. (1995). Earth system changes: Using environmental data for science teaching. *International Journal of Geographical and Environmental Education*, 4, 1, 107-115.

Fortner, R. W. (1996). Constructing Earth system science learning through multidisciplinary studies of global change. *International Geoscience and Remote Sensing Symposium, Proceedings, Vol II*: 1166-1168.

Fortner, R. W. and Boyd, S. (1995). Science is Understanding Planet Earth. II. Infusing Earth Systems concepts throughout the curriculum. Presented at the Annual Meeting of the National Association for Research in Science Teaching, San Francisco. ED 386 391.

Fortner, R. W., Pinnick, R., Shay, E., Barron, P., Jax, D., Steele, W. and Mayer, V. J. (1992). Biological and Earth Systems Science: Curriculum restructure from within. *The Science Teacher*, 59, 9, 32-37.

Fortner, R. W. and Mayer, V. J. (1997). Cooperative learning about global change: Prelude to cooperative action? *Proceedings of the Thessaloniki ([Greece] Conference on Environment and Society*, UNESCO.

Fulghum, R. (1993). *All I really need to know I learned in kindergarten*. Boston: Ivy Books.

Humphreys, B., Johnson, R.T. and Johnson, D.W. (1982). Effects of cooperative, competitive, and individualistic learning on students' achievement in science class. *Journal of Research in Science Teaching*, 19, 351-56.

Johnson, D.W. and Johnson, R.T. (1975/1994). *Learning together and alone: Cooperative, competitive, and individualistic learning*. Englewood Cliffs, NJ: Prentice-Hall.

Johnson, D.W. and Johnson, R.T. (1999). Making cooperative learning work. In Calderon, M. and Slavin, R., Eds. Building community through cooperative learning. *Theory into Practice*, 38, 2, 67-73.

Johnson, D.W., Johnson, R.T. and Holubec, E. (1998). *Cooperation in the classroom*. Edina, MN: Interaction Book Co.

Mayer, V. J., R. W. Fortner and Hoyt, W. H. (1995). Using cooperative learning as a structure for Earth Systems Education workshops. *Journal of Geological Education*, 43, 4, 395-400.

Mayer, V. J., Fortner, R. W. and Murphy, T. P. (1993). *Activities for the changing Earth System*. Columbus, OH: OSU Research Foundation with support from NSF.

Okebukola, P. A. (1985). The relative effectiveness of cooperative and competitive interaction techniques in strengthening students' performance in science classes. *Science Education,* 69, 501-9.

Slavin, R. E. and Fashola, O. S. (1998). *Show me the evidence! Proven promising programs for America's schools.* Thousand Oaks, CA: Corwin Press.

Slavin, R. E. and Karweit, N. L. (1984). Mastery learning and student teams: A factorial experiment in urban general mathematics classes. *American Educational Research Journal,* 21, 725-36.

Srogi, L. and Baloche, L. (1997). Using cooperative learning to teach mineralogy (and other courses too!). In Brady, J. B., et al, (Eds.), *Teaching mineralogy.* Washington, DC: Mineralogical Society of America, pp.1-26.

Stevens, R. J. and Slavin, R. E. (1995). The cooperative elementary school: Effects on students' achievement, attitudes and social relations. *American Educational Research Journal,* 32, 321-351.

CHAPTER 6: USING THE INTERNET IN EARTH SYSTEMS COURSES

William Slattery, Wright State University, USA
Victor J. Mayer, The Ohio State University, USA
E. Barbara Klemm, The University of Hawaii-Manoa, USA

1. INTRODUCTION

The Internet has become both a boon to teachers and parents and a source of serious concerns and problems. Both advantages and disadvantages are well known and publicized. However, those relating to teaching may not be, and the advantages in using the Internet for Global Science Literacy curricula especially need to be explored in some depth. In this chapter, we describe just a few of the uses of the Internet and suggest just a few of the problems. We believe that it is an essential learning environment for both teachers and students, especially for systems science based curricula. To aid in our discussion we have organized the uses according to the seven Earth Systems Education understandings (Figure 1, Chapter One).
The Internet provides the following types of assistance to teachers and students:
- Information on most aspects of science
- Illustrations to be used in teaching and projects
- Data bases to be used in science investigations
- News of current environmental and political events and issues
- A communication medium with other schools, students, teachers, scientists, political and business leaders
- On-line courses

We discuss examples of these uses under each of the understandings where appropriate. At the end of the chapter, we also point out some of the difficulties with the Internet's use and some of the cautions a teacher should exercise.

Never before have teachers and learners had the learning tools on hand that now make it possible for them to literally study the world/globe. In the past, at best learners engaged actively in studying local environments and passively (via texts, video), environments elsewhere. Now, the interactive multimedia learning tools integrated with the Internet open vast new possibilities.

Teachers/students can obtain images, text, graphics, sound & motion ON DEMAND, ANYTIME, ANYWHERE located via powerful new search engines. For example, now we can get satellite images created from untold thousands of data points measured by satellite instruments and interpreted by software designed to render the data in map-like formats. Much of what was once largely abstract and presented passively is now available for learners not only to learn from, but also to use actively in posing and seeking answers to their own questions. Classes which

93

V.J. Mayer (ed.), Global Science Literacy, 93–107.

were once self-contained and on their own can now network, to explore questions, to serve in host/remote quests for information, to gather, share, test data, etc.

We have not included the URL addresses of the sites discussed in the remainder of this chapter with one or two exceptions. We often find that addresses can be in error, or change. So instead, we have tried to give the exact title of the web page, or the organization that maintains it. Entering this information into a search engine will locate it. In the few cases where we tried this and it didn't work, we have provided the URL address.

2. IMPLEMENTING ESE UNDERSTANDINGS THROUGH INTERNET USE

2.1. Earth is unique, a planet of rare beauty and great value.

One of the most useful functions of the Internet for teachers and students is as a source of art and photography—images that can embellish any teaching or learning experience with the beauty of nature. Images of features such as flowers, erupting volcanoes, and animals are available. In addition, satellite images of tropical storms and other weather features can be obtained. Images of objects in space from the Hubble telescope and other astronomical instruments are spectacular. There are many public access sites maintained by government agencies where illustrations for discussions, projects and activities can be downloaded either into presentation programs such as *PowerPoint*, printed for making transparencies, photographed as slides, or just printed in hard copy. In doing this, of course, individuals need to be aware of and observe any copyright protections on the materials they are interested in using. With government sites, this is seldom an issue, but there are many other sites where one must be cautious, such as those selling photography. Normally use for teaching on a single use basis is sanctioned, but this is not always the case. To use the Internet materials a person must be careful to know and observe the source country's copyright laws.

The Smithsonian American Art Museum has over 5000 American paintings available as images on-line. In addition its web page has a vast collection of photographs accessible on-line. Mayer (1989) used images from this museum in an activity demonstrating the erosive characteristics of water in forming landscapes. He used the series of four images painted by Thomas Cole entitled *The Voyage of Life*. Students, in addition to learning aspects of erosion and landscape formation, also reported their aesthetic reactions to the images and the ideas concerning the stages of human life that they represented. The images used in the activity were actually purchased from the museum. Since then, we have had the advent of the Internet for use by the general public. Now, rather than ordering through the mail or actually visiting the museum in Washington, D.C. to purchase them, they can be downloaded while sitting at one's desk.

The famous Hudson River group of artists is well represented on this site. As well as Cole, the group includes Albert Bierstadt and Thomas Moran. Their paintings can be an excellent source of illustrations of American landscapes, as they existed in the 19th century. Comparisons of changes since then can be made by acquiring recent photos of the same localities to see what environmental changes have occurred. Students could provide an aesthetic analysis of the changes. Teachers in Japan might be interested in using a site of Sweet Briar College (http://www.artgallery.sbc.edu/ukiyoe/collection.html) that has images by Utagawa Hiroshige. He was a very famous 19th Century artist who used woodcut techniques in rendering Japanese landscapes. His most famous set of woodcuts was of the 53 stations along the Tokaido Road that ran from Edo (Tokyo) to Kyoto. He included views rendered in different weather conditions as well as many views of Mount Fujiama, a volcano.

Chris Whitcome, also of Sweet Briar College, maintains a site titled *Art History Resources on the Web*. Part 15 provides access to web sites with prints and photography. One of the sites available through his site has a number of Ansel Adams landscape photographs including his famous *Moonrise, Hernandez, New Mexico* taken in 1941. A site called Nature Photos On-Line (http://www.naturephotosonline.com/npo_main.htm) has images of animals, plants and landscapes. This is a commercial site so the use of the images is very limited. Its primary value would be for students to view selected images on-line, since they cannot be legally downloaded. In such a case, a teacher could construct an activity that would direct the student to certain photographs on the web site for analysis. Students (and teachers) could find it inspirational and aesthetically exciting to view nature through the eyes of these professional photographers and to study their interpretation of natural events and features.

An hour or two searching the web for 'art' and 'photography' sites can be very rewarding. A large array of sites available has images of value in providing an aesthetic context for teaching and learning, and for assessment.

2.2. Human activities, collective and individual, conscious and inadvertent, affect planet Earth.

As responsible members of the Earth system, students and teachers in GSL classes will be concerned about maintaining the environmental integrity of their habitat. Here, the Internet can provide a great deal of useful information and teaching materials. The greatest challenge facing the world community today is global warming. The Union of Concerned Scientists maintains a web page on Sound Science Initiatives. It has up-to-date and accurate information on global climate science and initiatives. For members of the site it has *PowerPoint* programs on the *Third Assessment Report* of the Intergovernmental Panel on Climate Change (IPCC) available for downloading. Members of the website receive periodic e-mail updates on governmental actions affecting climate change and other issues.

Another excellent source of current information on environmental issues is the *Environmental News Network*. Subscribers receive four to five e-mail messages a week on current environmental matters, from political and business actions, to weather, earthquake, and volcanic occurrences. Each of the short abstracts will take the reader to the web site for a fuller treatment of the topic. The United States

Geological Survey maintains a web site entitled *Earthshots: Satellite Images of Environmental Change*. Here teachers and students can examine satellite images taken of an area at different times, see a map of the region and read a description of the changes they can observe. Localities around the world are included—at least one on each continent. A professor at the Oklahoma Department of Geography maintains a site called the *Great Mirror* where images of culturally influenced landscapes are stored. Such images would be useful in demonstrating human impact upon the environment.

The World Resources Institute maintains a web site that would be invaluable for both teachers and students. This is one of the most respected organizations conducting research on the environment, population and other futures issues. Its Earth Trends page has a database, which can be searched either by variable (topic) or country or region. For example, a student could create data tables for Australia and its carbon emissions from bunker fuels over a series of years. Using this data, the student could develop time trends for the emissions from that particular fuel—or could compare its carbon emissions with those from gaseous fuels.

The National Science Foundation supports a web site called *Water on the Web*. It includes data on water quality, student lessons in how to use the data, and suggestions for teacher assistance. The data is in a format that is clear and easily used for examining problems covered in each of the student activities. Fish stocking decisions, diel temperature variation in lakes and thermal stratification are among the topics that can be investigated. GSL parts from traditional science programs in that there is an increased emphasis upon carrying out science investigations using systems approaches. To do this, the Internet will be an essential tool.

2.3. The development of scientific thinking and technology increases our ability to understand and utilize Earth and space.

This understanding emphasizes the need to include the broad variety of science approaches to studying problems in the environment. This will require significant changes in the ways that teachers organize their curricula. Because of the history of science and the science curriculum, documented in Chapter Two, the nature of the physical sciences has provided the basic structure for science curricula. This is because of the overriding need of national priorities for defense and economic competition. Perhaps another reason for the domination of school curricula by the characteristics of the physical sciences is the relative ease of data gathering and interpretation of physical science data in laboratory settings, and the continuing emphasis upon "hands-on" experiences in science. Students can easily collect data concerning, for instance, the refraction of light with very simple materials. Teachers of science have found it easier to develop student laboratories that are able to verify many physical science principles in unsophisticated school laboratory environments. In the past, this was not true of historical or system sciences. In order to generate meaningful data sets in system science, one must often be able to gather more than the local data formerly available to teachers of science. Now with the availability of data sets on the Internet, this problem can be solved.

2.3.1. On-line sources of data

In the research reported in Chapter Fifteen, a unit on typhoons was brain stormed by the teachers involved in the study. One of the activities they suggested was the tracking of a typhoon in the vicinity of Japan. Daily satellite images with current typhoon locations and predicted paths were available from the *Tropical Storms, Worldwide* web site. From knowledge of the characteristics of weather systems that spawn typhoons and how typhoons respond to varying conditions, students could predict their locations on subsequent days. Then on a given day, they could locate the typhoons on the satellite image and determine the accuracy of their predictions. Over time they could see that prediction accuracy varied with how far out in time they made their predictions. Another excellent source of weather data, called *OSU Weather*, is maintained by The Ohio State University Department of Geography.

The United States Geological Survey maintains a near real-time earthquake list available as a link through their National Earthquake Information Center home page. If students plot the longitude and latitude of earthquakes on a geophysical chart over the course of several weeks, the plate boundaries will be roughly outlined. Plotting deep and shallow earthquakes in different colors highlight the mid-ocean ridges and subduction zones. A similar exercise is possible by using the Current Volcanic Activity link on the University of North Dakota's *Volcano World* web site. This site has a great array of information about volcanoes. Plotting data from both the earthquake and volcano sites on a geophysical chart links the positions of deep earthquakes and most volcanoes to subduction zones. Knowledge built in this way extends students' understandings of the Earth system, and prepares them for more complex analysis of Earth systems data.

The National Oceanic and Atmospheric Administration's (NOAA) web site is a case in point. Their National Geophysical Data Center (NGDC) web page provides data for, in their words, Global ecosystem science and Global Change. Using *Geographical Information Systems*, *Database Management Systems*, and spatial modeling and inference tools, the global ecosystem database provides integrated research quality data sets. The database parameters are chosen for their potential use in integrated studies of climate, soils, human disturbance, and vegetation under present conditions and under global change scenarios. Geographic and political references, terrain, and other static data are provided for boundary conditions.

A site maintained by the Lunar and Planetary Institute is a source of extensive images of the moon and the various planets. Such can be used for studying variations in cratering on the different bodies, comparisons of landforms and other activities. NASA's Space Science web page provides images not only of planets, but also of many objects beyond the solar system. Especially useful would be those taken by the Hubble Space Telescope. Another site, the *Lunar Atlas*, provides images and locations covering the moon.

2.3.2. The GLOBE program

A model program for engaging students in the process of Earth systems science is the GLOBE Program. The acronym GLOBE is short for Global Learning and Observations to Benefit the Environment. Initiated by U.S. Vice President Al Gore on Earth Day (April 22) 1994, as a program that would allow students to make a meaningful contribution to the study of the Earth's environment, capitalizing on the Internet as a means for sharing information across the globe. The program was shaped through consultations among scientists, educators, and technology specialists representing academia, government, the private sector, and the non-profit community.

In March 1995, GLOBE students first began reporting data via the Internet to the GLOBE Student Data Archive (www.globe.gov/). In the first five years of the program, GLOBE students representing over 85 countries have reported over 4 million observations in the areas of atmosphere and climate, hydrology, land cover, and soils. GLOBE is managed in the United States through a federal interagency partnership of several science agencies and departments and was implemented through bilateral agreements between the U.S. government and governments of partner nations.

The GLOBE Program brings together K-12 students, teachers, and scientists from around the world who work together to learn more about the environment. Teachers guide their students through daily, weekly, and seasonal environmental observations, such as air temperature and precipitation. Using the Internet, students send their data to the GLOBE student data archive. Research scientists, such as Dr. Susan Postawko of the University of Oklahoma, are currently using the student data archive to investigate several aspects of Earth systems to begin to answer questions such as: How much water is evaporating from the surface of the Earth? How much water is in the atmosphere? How much precipitation is there in the atmosphere at any given time? Students in GLOBE schools around the Pacific Rim have been collecting data since 1997 to assist Earth systems research scientists such as Dr. Postawko. Another team of researchers is developing computer models to simulate and ultimately predict the behavior of ecosystems. GLOBE students assist this endeavor by collecting land cover data such as the percentage of canopy cover and vegetation type in a pixel sized plot of land. Each of their data collection sites is located by using Global Positioning Satellite technology. Researchers are then able to ground truth data collected from Landsat satellites.

Not all GLOBE Earth systems science research is carried out by scientists. The GLOBE program encourages research by and among its participating schools. Examples of international student research projects include Water Quality, Observing the Earth System, and Sustainable Development. An example of an international Earth systems project between schools in several countries is the GLOBE Weather Watchers project. Perhaps one of the most fluid and dynamic systems of Earth is weather. With resources provided by the American Meteorological Society, the Royal Meteorological Society, the Australian Weather and Oceanographic Society, and the newly formed European Meteorological Society, this project promotes sharing weather data between project members, and providing qualitative data on severe weather and/or weather extremes in different countries. The purpose of this project is to promote weather awareness, understand weather in the Earth system and to gain a global view of weather systems and their impact on the environment. The means to this end include the development of an international learning community, supporting collaborative student research, and interpreting real time data. Schools from fourteen countries are involved in the GLOBE Weather Watchers project. Students correspond via e-mail regarding their respective communities, their cultural and geographical setting, and share images of their GLOBE study sites. Students are able to ask questions about specific weather phenomena such as tornadoes in the United States, or typhoons in the Pacific.

GLOBE students are thus able to concretely make the link between the collection of data, forming working hypotheses, analyzing data, drawing conclusions, and dissemination of results. By collecting data, reporting it to an archived database, and using the data to assist researchers and develop their own research questions GLOBE has become a model for using the Internet to foster global scientific literacy.

2.3.3. A local and regional effort of data sharing

Similar efforts can be conducted on regional or local levels, using the Internet for the sharing of data. One was undertaken by the Departments of Geological Sciences and Teacher Education of Wright State University using Cedar Bog in west central Ohio as a source of data. Cedar Bog is a remnant of the visit of glaciers to Ohio. It is the southernmost alkaline White Cedar fen in the United States, an undisturbed area that now encompasses 460 acres, but once covered over 7000 acres. A unique set of geological and hydro-geological conditions maintain a delicate environment for many rare and endangered relict plant species. Small changes in elevation within Cedar Bog result in profound changes in the nature of the plant communities. Less than a one meter change in elevation within Cedar Bog can mean the difference between standing in a sedge meadow or in a grove of mature Tulip trees towering 25 meters over head. The upwelling of groundwater at an almost constant temperature of 12 degrees Celsius maintains the conditions needed to preserve this Pleistocene plant community. Cedar Bog is a popular field trip destination for school children, and a textbook example of an interdependent system. Scientists collect real time meteorological and biological data from Cedar Bog, and archive the data on a Wright State University K-12 Outreach homepage. Air temperature, humidity, wind speed and direction are some of the physical data collected.

Several species of animals including raccoons, squirrels, and white-tailed deer have been fitted with radio transmitters. Inquiry based field trip activities for children have been developed, and once back in the classroom, they can continue their investigations of Cedar Bog by accessing the physical and biological data on the Internet. By means of these data they can investigate questions such as; "Are deer more active on sunny or cloudy days?" and; "Is there a relationship between bird migrations and weather fronts?" Linking the physical experience of visiting Cedar Bog with the extended internet based investigation of the interrelationships between the physical and biological conditions provides students with a local sense of place, ground them in understanding local systems, and prepare them for extended investigations of global systems.

2.4. The Earth system is composed of interacting subsystems of water, rock, ice, air, and life.

2.4.1. Sources for basic science activities and materials
GSL curricula integrate all of the sciences. Concepts from physics, chemistry and biology are essential elements in developing the understandings of the Earth as a system. Thus, teachers and students will need to access information from web sites that provide instructional materials typical of those disciplines. A site with virtual activities on physical science principles is *ExploreScience.Com*. This is an excellent site for clarifying basic ideas such as the principles of wave motion, optics, electricity and magnetism, velocity, acceleration and inertia.

There are a number of sites that make available to teachers and students basic science activities and facts. The NSF supported the development of the Digital Library for Earth System Education (DLESE). It is a source of digitized Earth systems teaching materials for all educational levels. The goals of the library are:

> To develop collections of high-quality materials for instruction at all levels and covering all components of theEarth system.

> Provide access to Earth data sets and imagery, including the tools and interfaces that will enable their effective use.

> To develop discovery and distribution systems to efficiently find and use materials encompassed by the DLESE network.

> Provide support services to help users most effectively create and use materials in the DLESE "holdings".

> Through new communication networks to facilitate interactions and collaborations across all interests of Earth system education (Adapted from DLESE homepage; http://dlese.org/).

2.4.2. An on-line course for teachers
The Center for Educational Technology and Wheeling Jesuit University has developed on-line Earth Systems courses for elementary, middle school, and high school in-service teachers. The courses seek to develop content understandings of Earth systems science, model pedagogical methods suitable for the designated student age groups, and foster the use of Internet technology in K-12 classrooms. The middle school version of the Earth systems course uses real world events such as rain forest deforestation, volcanic eruptions, hurricanes, and sea-level rise as

platforms to achieve the course objectives. Teacher participants work on-line in teams during three-week learning cycles based on each event. Learning cycles follow the model of Karplus and Thier (1974) beginning with an exploration phase in which course participants interact with the material and with each other. This is followed by a concept development phase that asks them to work in groups to produce a report describing the impact of the event on the lithosphere, hydrosphere, biosphere, and atmosphere and the positive and negative feedback loops between the spheres. In the third and application phase, teacher participants develop Earth systems activities for their own classrooms, and post them to the course web site for other participants to review and comment upon. No formal textbook is used in the course, but instead participants use a variety of teacher sourcebooks and CD-ROM's selected to assist participants in developing content understandings as well as to assist in the creation of classroom applications. Before beginning the formal coursework, participants are given an introduction to Earth systems science, and a non-graded preliminary Earth systems study topic based on the fires that ravaged Yellowstone National Park in Montana, U.S.A. several years ago. The practice event is well documented and serves as a case study and model for the thought processes involved in analyzing Earth systems interactions. The following passages from the Yellowstone fires section show how these thought processes are built.

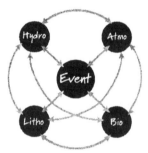

Figure 1. Earth System diagram.

When you look at the Earth System Diagram (Figure 1), you see that it has an "event" box in the center and that arrows run to each of the four spheres as well as away from them. Consider how that event might alter each of the systems. When you think of a possible "impact," write it alongside the arrow that leads from the "event" box to the particular sphere that is affected. The idea is for you to examine how this event might change or impact the different spheres that make up Earth's systems. When you do this analysis for another event or with your students, you can look at the spheres in any order you want. But for this example, consider how the Yellowstone fires might affect the spheres, beginning with the atmosphere.

Atmosphere. The fires created tremendous heat, developed fast rising columns of heated air, which in turn brought in more air at the base of the fires, dropped the humidity in the area of the active fire to near zero, and made dense smoke, which actually traveled hundreds of miles. There were certainly other impacts on the atmosphere, too. What ones can you think of that might be added to the list? Pick any two or three and put them alongside the arrow going from the event to the atmosphere.

Hydrosphere. What impacts might you expect the event to cause in this sphere? Is the addition of burning embers from the falling debris worth considering? Write your ideas alongside the appropriate arrow. Don't worry about getting them "all"--you probably won't be able to. Just focus on identifying one or two impacts. And remember, the impacts on some systems will be more numerous than on others. That's not a problem.

The course introduction includes similar questions regarding the biosphere and lithosphere. Emphasizing that the success of future investigations will depend on the collaborative nature of the interactions between team members, the course then asks participants to review the effects they anticipate in each of the spheres, asking themselves: "If this happened to the biosphere because of the fires, how might that have an effect on the fires themselves?" For example, if the fires cause the forest floor debris to be burned off in the biosphere, how might that impact the fires? Clearly, the fires would be removing fuel to maintain themselves. Again, class participants are asked to consider feedback interactions between the event and the other spheres. Finally, when students have a general idea of how the event and spheres can impact and affect one another, it is time to introduce one more level of interaction on the diagram.

Once the event has had an impact on a sphere or made a change in a sphere, for example the lithosphere, how does that influence the way the lithosphere now interacts with the other spheres? As an example, the fires burned the soil to a depth of several inches or more in certain "hot spots". What effect do you think that will have when it finally starts raining and there is runoff to the streams (hydrosphere)? If there were burning embers put into the atmosphere that eventually fell back into the stream (hydrosphere) how might that blackened debris have affected the aquatic invertebrates and fish (biosphere) that survived the event? In other words, look around at the "outside" ring of spheres in your diagram. If certain changes were created in one sphere by the event, how might those changes influence change in the other spheres? These interactions and interdependencies among the spheres can be written in along the lines on the outside of the diagram.

Teacher participants in the on-line class ponder these questions in threaded discussion areas, using the Internet to search for data to support their assertions, sharing their discoveries, and forming on-line learning teams to investigate the links between Earth systems, as they work through the different events used as springboards to Earth system science.

Following the course introduction and preliminary group work in Earth system science, participants engage in the study of four events in three-week learning cycles. Week One of each cycle is termed Sphere study. The purpose of sphere study is to learn how major events such as volcanoes, hurricanes, glacial melts, or deforestation might affect the spheres. The goal of each team is to present an analysis of the effects of a specific event on one sphere, identify the key concepts for the sphere and how they relate to the event. They become an "expert" in one sphere by working with others in their group to determine the event's impact on that sphere. By studying a different sphere for each event, they have a chance to develop their content knowledge of each sphere. The Internet is an integral part of the process. Students are asked to begin each weeks work by identifying any initial questions and prior theories they may have regarding the event and the interactions between the event and the sphere of the Earth their group is currently studying in a private journal space. Then they use hotlinks within the course to bring them to web sites that begin the process of learning about the interactions between the sphere and a given event, such as volcanic eruptions, rising sea levels, or destruction of the rain forest. Working individually at first, participants develop sphere to event interactions and use several different lines of evidence to support their position. Then in threaded discussion areas the groups work together to develop a description of the sphere, short and long term changes to the sphere as a result of the event, and feedback loops that may mitigate or enhance the effect of the event on given Earth subsystem. They post this effort for a group grade. Group products are graded by using a rubric for evaluation. By using a rubric, participants are able to become more comfortable with the distance learning process than they would be if evaluation criteria were vague. Rubrics are provided for all participant graded work. For example, Figure 2 contains the criteria for grading and the rubric for Sphere Space.

---

1. A description of the sphere being studied, with reference to key Earth science concepts.
2. A description of how the event may affect the sphere.
3. Identification of the short- and long-term effects and the destructive and beneficial effects of the event on the sphere.
4. Descriptions of feedback loops that could affect the event.

To earn five points:

1. Include key Earth science concepts in an accurate and thorough description of the sphere.
2. Describe in fine detail how the event could affect the sphere.
3. List more than one short-term change, more than one long-term change, more than one destructive effect, and more than one beneficial effect to the sphere.
4. Describe in thorough detail the positive and negative feedback loops that could affect the event.

To earn four points:

1. Include Earth science concepts in a correct description of the sphere.
2. Describe in some detail how the event could affect the sphere.
3. List one short-term change, one long-term change, one destructive effect, and one beneficial effect to the sphere.
4. Describe in some detail the positive and negative feedback loops that could affect the event.

Figure 2. Grading criteria and rubric for Sphere Analysis Activity

Rubrics with decreasing expectations set criteria for earning fewer points. In week two new groups are formed with different members who have studied other spheres the previous week. The purpose of event study is for each team to analyze how the event has directly affected specific spheres, how the affected spheres may trigger changes in the other spheres, and how the changes may influence the event itself. For each event the teams use the Earth system diagram to see the interrelationships among the spheres, and write a narrative that synthesizes the team's analyses of the event and its impact on the spheres. Teams may present their conclusions in Event Space in any way they choose. For example, they may tell the story of the event in terms of the changes in the spheres in the past, present and/or future, or describe the effects of the event in terms of the interrelationships of the spheres from the perspective of a human inhabitant. Each analysis must include a thesis statement, assertions that support the thesis statement, and evidence from readings, web sites, and experts to support each assertion. Although groups are allowed to develop any form of on-line cooperative work strategy to produce the event study, groups usually appoint an editor on a rotating basis. That individual will ask group members to submit a synopsis of their sphere study, and will assemble a first draft, available in the threaded discussion area. The entire group will then review the draft and suggest areas that may need to be strengthened. Individuals will search Internet data and resource sites, as well as printed material to flesh out the details of the interrelationships between the event under consideration and the atmosphere, hydrosphere, lithosphere, and biosphere, and among the spheres themselves. The draft is revised, and a final group product is submitted. As with Sphere study, the submission is graded by using a rubric assigning higher points for powerful, comprehensive statements in the thesis statement, and using multiple lines of reasoning to support their assertions.

During week three of each three-week study cycle, individual participants will apply their knowledge of Earth system thinking gained from Sphere Space and Event Space, to devise activities to be used in their own classrooms. In Classroom Application Space, participants are asked to describe how they can apply what they learned into teaching practice. These classroom applications are individual efforts devised to demonstrate how the experiences with their sphere group and event team can be made relevant to K-12 students. For example, they might share an Earth Systems Science activity they tried with their students and their reflections on how it went.

At the end of each event, students have an opportunity to earn extra credit by completing a local event system analysis by discovering a local event and discussing how that local event affects the four spheres. An example might be the creation of a dam on a local river or the impact of a corporate hog farm on a local community. Performing a local event system analysis is an individual project and each local event submission must contain a thesis statement, assertions that support the thesis statement, two sphere effects, and evidence from reading(s), web sites and experts to support each assertion. Students complete a journal entry each week for the duration of the course. This is a private space where they reflect on their personal growth and community relationships. The goals of journal writing in the course are to track progress in developing Earth system concepts.

The on-line course concludes with a final project, designed to evaluate the student's content understanding of Earth systems science as a result of having taken the course. The final project is an event study performed individually. Students are given an event, and asked to analyze the event in terms of the overlapping and interacting effects on the spheres. Post course participant assessments reflect an increased awareness and knowledge of Earth systems science and expanded confidence in using both the course pedagogical methodology and telecommunications technology in their own K-12 classrooms. Teachers are thus armed with the information needed to develop scientifically literate students. Yet developing student understandings of global interconnectedness requires more than K-12 educators able to use technology to bring data and research information into the classroom. It requires the hands-on participation of K-12 students in the scientific process.

2.5. Planet Earth is more than 4 billion years old and its subsystems are continually evolving.

At present the authors are not aware of any materials on the web that would assist learners in gaining a sophisticated understanding of "deep time", or for that matter, the extreme distances involved in the next understanding, although there is a site, *Dating Techniques*, that describes the various ways in which scientists date materials. Trend, in Chapter Thirteen reviews the research that has been conducted on developing of the ability to understand vast expanses of time. He offers some suggestions to teachers.

However there are resources that would add context to any investigations of concepts associated with deep time. For example, evolution is well represented on the web—both from the scientists' evidence and conceptualization and from the creationists' perspective. The National Academy Press has the National Academy of Sciences, *Teaching about Evolution and the Nature of Science*, on its web site. It is directed toward teachers and provides reasons for teaching evolution, major themes to be developed, frequently asked questions, and some activities on how to teach the concept.

Britannica.com has a great site titled, *Discovering Dinosaurs*. It provides a complete history, with visuals, of the discovery of dinosaurs and the development of theories about their morphology and ecology. The University of California, Museum of Paleontology, has a large web site with virtual exhibits on Evolution, the Diversity of Life Through Time, and Geologic Time. *BBC Education* has an *Evolution* web site. It has Darwin's complete *Origin of Species* on the site as well as other features, including a simulation, *Biotopia*, in which students can create an ecosystem with creatures and see it change over time.

2.6. Earth is a small subsystem of a solar system within the vast and ancient universe.

With this understanding, once again, we know of no materials or ideas on the web that would help students develop an adequate conception of the great distances involved in the universe. There are excellent sites mentioned earlier in this chapter, however, that are maintained by NASA and provide outstanding visuals.

Students for the Exploration and Development of Space (SEDS) have a web site that includes a wealth of features, including a space images archive, a solar system tour, the *Messier Deep Space Catalog*, and astro maps. This is a good resource for students investigating characteristics of space objects.

2.7. There are many people with careers that involve study of Earth's origin, processes, and evolution.

Many of the sites already mentioned can be used by students to study the type of work carried out by scientists and technicians. But the teacher should not limit the examination of the role of Earth systems in students' future lives to this. In avocations, such as photography, skiing, swimming, boating, any avocation one can imagine, in some way or another involves the Earth system. Therefore, web searches can identify numerous sites where a career in science and technology, recreational services, or simply an avocation can be observed and studied. Using web sites a student can develop an understanding of that particluar career he or she is interested in. The teacher can also capitalize on students' interests in careers and avocations and link them to the study of Earth processes, using a variety of Internet sites.

3.   CAUTIONS WHEN USING THE INTERNET FOR GSL COURSES

There are a number of concerns and cautions teachers and students need to observe when using the Internet. The most obvious that has been found by teachers during student use of the net is to ensure a firewall between the student and pornographic sites. Software now available is not 100% effective, thus the teacher must carefully supervise student use of in-school computers connected to the net.

Each science and environmental site needs to be carefully evaluated for the accuracy of content. Those that are maintained by federal science agencies, such as NASA and the USGS, can be relied upon for scientifically valid information. Sites maintained by industrial consortiums, pseudo-scientific organizations, and others with political messages, such as the creationist groups, can often be used by students and teachers for purposes of comparing their content with that from legitimate science organizations--but they should not be relied upon for valid scientific information. Often the sponsors will be hidden by names that sound scientifically legitimate but in reality they have a political agenda. There are several, for example, that focus on critiques of the science of global climate change.

If materials are downloaded from the Internet, the teacher and student need to be careful to observe any copyright protections that have been applied to the materials. And finally, the Internet should not be used as simply another purveyor of information. It allows creative ways for students and teachers to share information across great distances, to collect data from sites in other countries, to locate effective learning and assessment materials, and to find images that can assist in creating new and interesting materials.

4. CONCLUSIONS

To predict the future use of the Internet to foster global scientific literacy is an impossible task, but the avenues to that future are being built today. Programs such as GLOBE, the Earth Systems Science on-line courses developed by the Center for Educational Technologies, and the many and varied data archives maintained by government and institutions are paths down which future program and course designers will travel to use the Internet as a bridge to Global Science Literacy. There are many agencies, institutions and individuals who have gone unmentioned in this chapter. They labor each day to further the goal of global scientific literacy by developing on-line courses, maintaining on-line data archives, and mentoring students and colleagues in K-12 education.

REFERENCES

Karplus, R. and Thier, H. (1974). *SCIS Teacher's Handbook*. University of California, Berkeley Ca: Science Curriculum Improvement Study.
Mayer, V. J. (1989). Earth appreciation. *The Science Teacher*, 56.3, 60-63.

CHAPTER 7: DEVELOPMENT OF CHARLES DARWIN AS AN EARTH-SYSTEMS SCIENTIST: A FIELD EXPERIENCE

David B.Thompson, Keele University, UK
Modified by Victor J. Mayer, The Ohio State University, USA

This chapter is based in part on the author's article originally published in *Teaching Earth Science*, Volume 24, Number 2, 1999, entitled "Charles Darwin's Presence in North Staffordshire C. 1815-1842: The Rationale for some Educational Initiatives".

1. INTRODUCTION

Field experiences are essential in GSL courses. In the 1988 conference discussed in Chapter One, many of the scientists indicated their interest in science was stimulated by the beauty they found in nature. This fact provided the basis for formulating ESE Understanding One (See Table 1, Chapter One). So too should children able to experience the beauty of nature in their science courses. In GSL we emphasize the importance of incorporating the aesthetics of the Earth in science classes, perhaps not as a teaching objective, but at least in providing a context for learning. Certainly taking students into the field can provide that context. If in addition the teacher can provide experiences similar to those experience had by a famous scientist—field experiences that influenced that scientist to take on science as a career—then students will learn a bit more about why people become scientists and thus understand more about science itself.

Charles Darwin was a giant of science. Yet much of his life was centred in the vicinity of his childhood home, the Keele area of Central England. In describing his early life in this chapter and relating it to the environment in which he lived, we provide a vivid description of a scientist of the 1800s and the influences on his life and his profession. In addition, this chapter and the exemplar of an excursion schedule it includes are intended to:

- Highlight the need for science teachers to make the most use of the opportunities provided by the local environment when conceiving courses which aim to contribute to a holistic general education.
- Encourage science teachers to research and use the lives of "local or national heroes" (to use the phrase of the popular BBC TV programme of Adam Hart-Davies) as a medium for enthusing pupils of both genders when planning schemes of work.
- Promote environmental Earth-systems education locally out-of-doors in the natural environment, to which all pupils might have access at some time in their school career.
- Illuminate and highlight on-going science-based issues in the district or region around the school. In this case, by way of illustration, the problems

109

V.J. Mayer (ed.), Global Science Literacy, 109–128.
© 2002 *Kluwer Academic Publishers. Printed in the Netherlands.*

are centred on local planning applications in two small areas of Central England. In all these instances, the understanding of the scientific background does not change very much over time. However, the outcomes of the planning applications depend very much upon the acceptance or denial by the planning authorities of changing value positions regarding both the conservation and development of the area and the tenor of public opinion at the time.

• Bring these matters, and the general principles lying behind the development of any exercises relating to them, to the attention of an international audience rather than that for which the exercises were originally developed in a local area.

The rationale behind these aims is that, within 100 kms of any school, there may have lived "local or national heroes" (i.e. famous scientists, technologists or other persons of note). It is the professional task of teachers and schools to capitalise upon the lives and contributions of such persons for the common good of society in order to inspire and generally educate their pupils in a cross-curricular way.

However, bearing in mind the international audience that is addressed in this book, Table 1 suggests some of the historic and more recent Earth systems scientists in different countries who might be considered as "local or national heroes" in the present educational context.

Table 1. Scientists who might be considered to qualify as "local or national heroes" in the context of contributing to the understanding of the Earth as a System.

Country	Historical Earth-systems scientists	Modern Earth-systems scientists
Australia		Sam Carey
Canada		Tuzo Wilson
China (Switzerland)		Ken Hsu
England	William Buckland, Alfred Wallace. Charles Darwin, Arthur Holmes	Fred Vine, Dan McKenzie, James Lovelock
France	Georges Cuvier, Nicholas Desmarest	Xavier Le Pichon
Germany	Alexander von Humbolt, Abraham Werner, Leopold von Buch, Alfred Wegener	

Italy	Galeleo Galelei	
Japan	Kenji Miazawa	Hitoshi Takeuchi
Scotland	James Hutton, John Playfair, Charles Lyell	
South Africa		Alex. du Toit, Lester King
Sweden (France)	Carl von Linne' (Linneaus)	
United States	J.K. Gilbert, William Morris Davies, Harry Hess, John Wesley Powell, Clarence King	Eugene and Carolyn Shoemaker, Barbara McClintock, Walter Pittman, Anita Harris, Stephen Gould, E. O. Wilson, Jack Horner, Lynn Margulis

The story of the development of Charles Darwin as a scientist in such a global context is salutary, for he grew up to command the knowledge, understanding and approaches of many disciplines relevant to Earth Systems science. In those days, most of these were subsumed under the general title of "Natural History". His interests included:

- **as a junior and teenager**: minibeasts (e.g. insects; especially butterflies and beetles), all kinds of plants, fossils, minerals and rocks (he wrote that he wished to know the origins of every pebble!); chemistry and gardening;
- **as a university student**: at Edinburgh and Cambridge: botany, zoology (especially entomology); ecology (especially of the seashore); mineralogy, chemistry; theories of transmutation with descent (both the ideas of his grandfather Erasmus Darwin and Jean Baptiste de Lamark); the history of the Earth (the Huttonian view of the workings of the Earth as a heat engine, involving the emplacement of plutons and the action of volcanoes, versus the Wernerian-Jamesonian view that all rocks were precipitated from a universal ocean; see later for further detail);
- **as a global voyager**: oceanography; flora and fauna, and their bio-geographic distribution in the oceans and continents visited; ecology and palaeontology/palaeoecology/palaeogeography; vulcanology, petrology and petrogenesis; metallogensis; seismology, structural geology; glaciology, geomorphology and anthropology of the places visited, from uplifted plains to mountain ranges and coral islands;
- **as a maturing scientist at home**: after 1836: the consideration of all of the above as an author, together with: the Linnean affinities and classification of barnacles (1851-4); ideas on transmutation and evolution (1859) and,

increasingly, an understanding of aspects of science where evidence could be derived from practical pursuits or even hobbies: the breeding of domestic animals and pigeons (Darwin, 1868), the fertilisation of orchids by insects (Darwin, 1862); the generation of new strains of crops and the principles of improving agriculture generally involving the origins and fertilisation of soil and the life history and functions of worms (Darwin, 1881) etc.

In relation to all the above aspects, he was a voracious reader of scientific literature and the philosophy of its methodology. He corresponded with all the principal investigators of the day and drew many of them like Thomas Huxley (polymath) and Joseph Hooker (botanist), into day-long and weekend-long seminars at his home in Downe, Kent, in which his latest experimental results, his emerging ideas and the opposing views of the day were thoroughly examined.

He taught himself the virtues of combining **experimental, deterministic approaches** to problems (in the investigation of botanical speciation - cross pollination, hybridisation etc. (Darwin, 1862, 1876), the roles of worms in generating fertile soil (Darwin, 1881), etc. with **historical, descriptive approaches** wherein scale, rate and time could not be modelled and it was impossible to set up or control variables either singly or in concert. In these latter cases, he understood the need to retrodict distant or recently past events, as for example recorded in rock sequences (Darwin, 1839, 1842, 1846), based on understanding present-day processes and applying likely rates of energy transfers. So far from being a Baconian inductivist and following the preferred scientific methodology of the day (Whewell, 1837), in which an observer was expected to clear his mind of all hypothesis and theory and observe only the facts (without bias or hint of preferred interpretation), he was a "philosopher" who conjectured and hypothesised constantly and tested a multitude of likely and unlikely ideas in a hypothetical-deductive, falsificationist way. In all these endeavours, he was pre-eminently a vividly imaginative and creative person.

In the section which follows, space does not allow the nature of Darwin's investigations on HMS Beagle (1831-36) to be recounted in detail. It is sufficient here to point out that during the whole of this voyage he worked in a thoroughly interdisciplinary context and sought to test a variety of competing, and often long-standing, approaches to investigating the workings of the Earth as a system, the understanding of its processes past and present and its feedback mechanisms. Here attention is drawn to the Beagle's investigation of the following broad aspects of the Earth system:

- **the atmosphere**: observing and recording a wide variety of weather phenomena at sea at set intervals, and their use for predictive purposes, was vital to the success of any sea captain;
- **the hydrosphere**: monitoring the strength and direction of ocean currents and their salinity and temperature, and applying known theory of their origins, was equally vital;
- **the biosphere**: recording the flora and fauna of the oceans and land areas, suggesting their field classification, ensuring the curation and transportation of key specimens to reputable museums back home, were basic to the

growth of global understanding. The receipt of such materials served to astonish and inspire the metropolitan audience and the potentially expert investigators of the specimens and to enhance the reputation of the young explorer to an inordinate degree;

- **the lithosphere**: the accurate surveying and plotting of the positions of islands and coastlines in terms of their latitude, and particularly longitude (Sobel and Andrewes, 1998), was also basic. The discovery and reporting of giant fossils like *Megatherium* (actually a giant armadillo later named *Hoplophorus*), and others, like the *Mastodon* (of equally startling novelty), were received back home with great anticipation and excitement. Rocks of alien type from novel geological environments like the Andes opened the eyes of the ruling elite to the recent origins of specimens once associated only with the basements of ancient mountain ranges in Europe.

Above all, in performing these duties, Darwin sought to test the applicability of many competing, and in some cases long-standing, approaches to the elucidation of Earth history and to establish how the Earth had developed and was functioning holistically as a planet. Ideas and methodologies associated with the following persons and theories were in contention and ferment in his mind as he sailed off in the Beagle in December 1831:

- Biblical scholars: Creationism (Anon, 1603), the idea that the Earth had been originated by a divine creator in six days, as set down in Genesis;
- Scriptural geologists: The idea that superficial deposits, now attributed to the Ice Age and subsequent river and wind regimes, were in some way related to biblical accounts of Noah's flood.
- Armchair theorists like Erasmus Darwin (1789; 1791), Charles' grandfather who, like many before him, constructed cross-sections of the Earth depicting the nature and workings of its interior and surface, largely from the comfort of his study and garden.
- Early evolutionists like Erasmus Darwin and Jean Baptiste de Lamark who speculated on the change of life forms with descent, the latter championing the inheritance of characteristics by offspring, which had been acquired by parents during their recent lifetime.
- Inductivists like Francis Bacon who believed that scientists could, and should, be unbiased collectors of fact whose interpretation would be obvious.
- Travellers like Alexander Humbolt (1814-29) who assiduously assembled details of Nature in distant continents.
- Abraham Gottlob Werner, the Neptunist, and his disciple in Edinburgh, Robert Jameson, who had tried to teach Charles that most rocks, including those that we now describe as igneous and metamorphic, were precipitated from a universal ocean which had since retreated to the interior of the Earth.
- James Hutton, the Plutonist and Vulcanist, whose disciples professed the idea (1785; 1795) to Charles and fellow students that the Earth was a machine driven by internal heat. He saw that its surface processes could be observed at the present day wearing away the land, depositing large

volumes of sediment in the sea, from which mountains and worn-down land areas would eventually be built up in an ever-repeating cyclic processes. Hence, Hutton taught that the rock record could be interpreted in terms of unimaginable lengths of time. Indeed, he discovered geological time, now popularly called "Deep Time", and stated that there was "no vestige of a beginning; no prospect of an end".

- Robert Hooke, James Hutton, John Playfair, Charles Lyell; Uniformitarianism, roughly the idea that "The present is the key to interpreting the past"; the suggestion that knowledge of the processes acting today, and arguably their rates, are the only means by which the rock record, and hence the history of the Earth, can be deciphered. Darwin received and avidly read copies of the three volumes of Lyell's *Principles of Geology* (1831-3) at various ports of call on the voyage of the Beagle. He continuously tested the validity of Lyell's methodology and registered considerable approval.
- George Cuvier (1813), Adam Sedgwick, Roderick Murchison: Catastrophism, the idea that the rock record contains evidence of episodes of unnaturally violent events. These were variously interpreted, but some events were associated with the intervention of the Creator, whose intention was to warn and punish sinful man.

In presenting the rest of this account, the author will concentrate on revealing some of the wellsprings of Darwin's creativity, as instanced in his formative development in an educational context (1809-42). The following sections will illustrate how teachers could, and should, lead visits to the still extant localities where such ideas were nurtured. They will show how such visits can help to stimulate the educational process for students of all ages. The outlines of two such field excursions will be presented later. Should readers wish to follow the second of these routes in greater detail they are referred to a preceding account (Thompson, 1999).

2. THE EVOLUTION OF CHARLES DARWIN AS AN EARTH SYSTEMS SCIENTIST

In modern parlance, Charles Darwin (1809-1882) is thought of as a "biologist" and only recently has his development as a geologist - the role he wished above all to assume in his early days - been of interest. He was born at the family home, The Mount, on the Welshpool Road out of Shrewsbury. In truth, we might now consider him one of the very early and most influential of the Earth system scientists, developing concepts with implications across several different Earth systems.

The following accounts of Darwin's early life are based upon a vast array of excellent sources which have been made available in recent years, but principally the following: Charles Darwin's autobiography, alas written as a relatively old man who was ostensibly writing for the benefit of his children but subconsciously perhaps had posterity in mind (see De Beer, 1974); the biographies of Desmond &

Moore (1991) and Browne (1995), and the commentaries on Darwin's life and times by Kohn (1985) and Bowler (1990).

His father was Robert Wareing Darwin (1766-1848), a doctor, moneylender and accountant to the Wedgwood potteries. He was the son of Dr. Erasmus Darwin (1731-1802), of Lichfield, an early evolutionist. His mother was Susanna Wedgwood (1765-1817), the daughter of Josiah Wedgwood I of Etruria Hall, now part of Stoke-on-Trent.

A signal event in Charles' young life was the death of his mother when he was aged eight. After this event, his father became notably lonely and morose and was prone, amongst other overbearing and autocratic things, to subject his children to a monologue for an hour or more before evening meal. Despite Charles' undying profession of love and respect for his father all his life, the children learned to look forward to weekends and holidays when they could escape the oft-times oppressive atmosphere of the Mount. Hence as a small boy and later, Charles frequently enjoyed visits to the home of his cousins and uncle, Josiah Wedgwood II at Maer Hall in North Staffordshire. The Hall was soon dubbed "Bliss Castle" by young Charles, a reference to its then somewhat castellated appearance and a part of the building which has since been dismantled. Occasionally he and his sisters visited other friends in the district, such as the Tollets of Betley Hall (who ran an improver's farm, Model Farm, now part of Betley Old Hall Farm) and the Sneyds of Keele Hall (the ramshackle old hall, built c.1580).

It was at Maer Hall that Charles learned to hunt, shoot, fish, entomologise, and further develop his taste for natural history. He also learned the delights of socialising with his sisters, their cousins and the interesting visitors, many of who were the foremost minds of the day, in a house described by Sandra Herbert (1993) as home of liberality and open-mindedness.

2.1. Schooldays

As a small boy in Shrewsbury, Charles had learned, and was learning, to collect all manner of plants, minibeasts, minerals and fossils and was encouraged by persons like "old Mr. Cotton". He showed interest only in possessing, naming and ordering his specimens - not in understanding their deeper nature and origins. This behaviour accords well with that recorded in psychological studies of the salient events which give rise to career choices of many scientists in later life - many steps are taken in primary school well before the age of ten (see Head, 1985, Woolnough et al, 1997). Charles was particularly attracted to mind-blowing theories which had to be generated to account for single accurate observations e.g. the presence of a Lake District igneous rock in the centre of Shrewsbury, outside of Morris Hall, with no present-day river available to transport it from its place of origin to the town; the identification of a single fossil shell of undoubted tropical affinity in the sands and gravels of the Drift close to the town. He was later puzzled, even horrified, that Adam Sedgwick appeared not to be interested in this latter anomaly when he pointed it out to him in 1831. Sedgwick is reported to have said: "If the shell was genuinely embedded there it would overthrow everything that was known about the superficial deposits of the Midlands Counties" (then considered to be Diluvial deposits DBT).

Sedgwick's hypothesis was that "It must have been thrown away by someone into the pit". This illustrates that scientists do not easily ditch their favourite explanations when a serious anomaly arises; they often invent auxiliary hypotheses to protect their ruling theories.

Charles spent his schooldays with his brother Erasmus (1804-1881) in Shrewsbury, mainly at the Free Grammar School. Neither boy was good at classics - the curriculum of the sons of gentlemen to be. His headmaster, Archdeacon Dr. Samuel Butler, failed to see in him any evidence of genius. Indeed, Butler upbraided both boys severely in private and in public for their interest in chemistry and natural history. The eminent theologian had heard of the undesirable smells which emanated from the Darwin boys' experiments in their garden shed when they followed the practices recommended in William Henry's book (1819) "Elements of Experimental Chemistry" - hence Charles's schoolboy nickname "Gas". Progress of both boys was slow. Charles was believed by his father to be in danger of becoming a wastrel and a "good for nothing" and was withdrawn from school. Both boys were regarded as "failures". (The concept and implication of "failure" at school lies deep in the English psyche! Thereafter, throughout his life and possibly as a consequence, Charles always seems to have lacked confidence in himself and in anything he was doing).

2.2. His introduction to Natural History

In 1825, at the tender age of 16, Charles was sent to follow his father and grandfather at Edinburgh University, that northern outpost of the Age of Enlightenment. There he joined his brother (recently transferred from Cambridge) in order to study medicine. The curriculum of the first years of the degree was commendably wide and depended on student choices and the payment of fees (shades of the present-day debate!). Charles was able to attend the courses given by Professor Thomas Charles Hope (1766-1844), a flamboyant lecturer and demonstrator in the Department of Chemistry and Mineralogy, and by Robert Jameson (1774-1854) of the Natural History Museum. The latter was a disciple of the world-famous Abraham G. Werner of the Mining Academy at Freiburg. Jameson was a dry old stick who lectured monotonously from notes based on his own books. Both men professed to study the same natural world, but Hope, attracting over 500 students each year, espoused the traditions of the experimental and physical sciences and favoured Huttonian Vulcanism-Plutonism and the viewing of the Earth as a machine driven by internal heat. Alas! Hope had given up researching in order to teach in extravagant ways and he never had his students perform his, or their own, experiments as did his illustrious predecessor Professor Joseph Black (1728-99). By contrast, Jameson, still favoured by as many as 250 students per year, adopted the Linnean-Wernerian approach, whereby Nature was to be investigated in order to reduce it to orderly hierarchical relationships and ranks of classified objects set out in museum cases. Fortunately, however, Jameson encouraged students to handle and investigate minerals, rocks, fossils and modern organisms for themselves in the backrooms of the museum using the simple tools and tests of the day (e.g. goniometers, physical and chemical tests). In addition, he encouraged them to attend and contribute to the student Plinian Society (founded 1823) and the Wernerian Natural History Society (founded 1808). He had them read original work as published in journals like the Edinburgh Philosophical Journal (of which he was the editor). Between them, Hope and Jameson's contrasting beliefs and styles made the study of the natural world one of abiding conflict and interest - a prime example of the importance of "creative tension" in promoting effective science education and, indeed, the very progress of science itself.

It was in these circumstances that Darwin further prospered from meeting Robert Grant (1793-1874), a Lamarkian evolutionist, and from going on field trips with him. These were usually to the intertidal zone of the Firth of Forth, during which time Charles debated the theoretical issues of the day, learned the rudiments of personal research and gathered data for his first paper. This he delivered nervously but triumphantly to the student Plinian Society at the tender age of 17. He also learned from Grant, alas, of the tendency of some scientists to develop professional jealousy and an over-acute sense of property and priority. This arose when Grant appropriated a discovery made in his "own" field by his young protégé and paraded it in print, albeit without a proper degree and place of acknowledgement.

Darwin later claimed that Jameson's teaching put him off geology for life, but Secord (1985) has suggested that this was not true; the annotations of Charles' textbooks and notebooks (extant in the Darwin library at Cambridge) show that he must have learned much of a practical and theoretical nature from the rich scientific learning environment that Jameson provided. (Indeed, the best schools today always attempt to provide just such an ambience; see Woolnough et. al, 1997). In particular, when he was on field excursions with Jameson, Charles learned to believe only his own eyes and not be taken in by forced interpretations of phenomena e.g. the origin of the dolerite dykes cutting the Carboniferous strata on Arthur's Seat and Salisbury Crags in Edinburgh. Jameson described these dykes as precipitates from a universal ocean that were emplaced within pre-existing *neptunian* fissures. Having students learn from the negative is a powerful mode that is much used by the present writer!

Sadly, however, the study of medicine proved to be both dull and shocking for the Darwin sons, for in successive years they were unable to stomach the trauma of witnessing operations involving the amputation of the limbs of children without anaesthetic. Hence, both withdrew from the University, Charles having learned providentially that there was a safety valve known as Natural History (which very much included geology) lying beyond any further aspirations that his family might press upon him.

2.3. His education as a potential Church of England minister

Charles returned home after two years at Edinburgh to discuss his future with his widowed father and, despite grave religious doubts, agreed to go to Cambridge in 1827 on a 400 Pound per annum allowance to study theology with the intention of entering the ministry of the Church of England. There he soon entered into the spirit of the exclusively male student life and became an "idle sporting man" at Christ's College. Fortunately, his interests included field excursions on horseback in which he was able to extend his studies developed at Maer and Shrewsbury and collect further specimens, particularly beetles and butterflies. He took a great interest in the efforts of the established churchmen to rationalise geological and other dangerous scientific findings in terms of the revealed world of an omniscient creator and he pored over William Paley's *Natural Theology* (1802). In his leisure time, he managed to avoid the attentions of Rev. Adam Sedgwick (Woodwardian Professor of Geology since 1818) who was the chief proctor responsible for policing the male students' riotous behaviour whilst the latter were carousing about the town. Charles claimed never to have attended Sedgwick's lectures. However, Secord (1985) has found evidence of Charles' discussions of Sedgwick's lectures, perhaps at second hand, and his study of certain aspects of geology, before he left the university. In the event, Charles graduated somewhat early, with only a <u>pass</u> degree, which placed him 10th out of 178 candidates. Thereafter he was required to spend two terms in residence, during which time he **may** have attended some of Sedgwick's annual course of 36 lectures. He was no doubt stimulated by the latter's approach to geology as a catastrophist and as a natural theologian who was implacably opposed to the uniformitarian and rationalist approach of Hutton and Lyell. Thoughts of Charles' future career were undoubtedly generated by his stay at Cambridge, for his tutors thought very highly of his promise as a naturalist. In particular, Charles had made a deep impression on the Reverend John S. Henslow (1796-1861), his general tutor of many years standing. Henslow had produced the definitive work on the *Geology of Anglesey* (1818) and had formerly been the Professor of Mineralogy (from 1822), but by that time (1825), he had been translated miraculously to be Professor of Botany. Indeed, the only formal science that Darwin undertook was two years of Henslow's course in botany. Darwin soon became known as "the man who walks with Henslow" and was held in some awe by his tutor who once remarked "What a fellow that Darwin is for asking questions!" Teachers will no doubt recognise these very desirable attributes which they aspire to develop in students and will recall how wearing it can become if they are successfully achieved!

2.4. An apprenticeship in field geology with Professor Sedgwick

In 1831, Charles returned home to Shrewsbury to discuss his future yet again with his father, but this time as a potential vicar of a rural parish. In order to spend his summer profitably, Charles had accepted a suggestion of the Professor Sedgwick that he should make a geological map of the Shrewsbury district. He bought appropriate instruments, practised measuring and plotting dip and strike in his bedroom, and travelled to Llanymynech to start work. Alas! he confessed that the task proved to be too daunting. It has been recently shown (Roberts, 1997) that Charles had trouble with his spelling and may have been dyslexic. He recorded data in his field notebooks which, like those of many a modern student, were 180 degrees from the true readings. Help was at hand, however, for Sedgwick had suggested that Charles should accompany him that summer on part of his field season in North Wales, to which end the former would find it convenient to meet him in Shrewsbury, stable his horse at the Mount and stay the night. During these few hours, Sedgwick quickly won the swooning admiration of at least one of Darwin's sisters and was equally rapidly diagnosed as a hypochondriac by Dr. Robert, Charles' father! Thereafter for five weeks, Sedgwick was able to tutor Darwin personally in making and plotting geological observations and to induct him into current ways of geological thinking. Very soon, in the Vale of Clwyd, Greenough's (1820) mapping of the Old Red Sandstone beneath the Carboniferous Limestone was seen to be amenable to an alternative interpretation - the rocks could be New Red Sandstones faulted down to a position topographically beneath the older limestone. Hence, Sedgwick argued that the Vale of Clwyd had originated by stretching and had sunken in order to accommodate the deposition of the New Red Sandstones - a quite prescient interpretation that hints at the presence of the initial stages of a modern Mackenzie-type extensional basin. Again, Charles found this kind of reasoning, and the overturning of established ideas by reference to a few very acute and accurate observations, to be much to his taste. Note, however, that modern historians of science caution teachers and students not to evaluate ideas from previous ages in any way other than in the historical context of their times.

2.5. The voyage of the Beagle; the fashioning of a Lyellian disciple

Upon return to Shrewsbury, Charles' discussions with his father resumed, but they were interrupted by a letter from Henslow notifying Charles of an opportunity to sail as a gentleman companion and naturalist with Captain Robert Fitzroy on HMS Beagle. This was a 90ft-long brig, which had been retained by the navy after a successful surveying voyage in order to chart the commercial shipping lanes of South America and parts of the Pacific Ocean over two years. In truth, Charles was fourth choice for this post (after Henslow, Robert Jenyns and one other). At first, his father scoffed at the idea of Charles embarking on a career as a naturalist, but later he relented and agreed to fund the venture after Charles had made several visits to Maer in order to enlist the support of uncle Josiah Wedgwood II. The letters recording the initial rebuttal of this wild idea by Dr. Robert Darwin (Darwin, R. 1831a,b) and the absolutely crucial interventions by Uncle Jos II before the final acceptance of the offer by Charles (Wedgwood, 1831), are held in the Wedgwood Archives in Keele and Cambridge University libraries respectively. By such delicate threads hang the career paths of even our most gifted students.

Thus, it was that in October 1831 Charles gathered together appropriate instruments and books and made haste to Plymouth to await orders to sail. Meanwhile he read and re-read the books of Humboldt (1814-29) and Herschel (1830) before he received an early copy of volume 1 of Charles Lyell's *Principles of Geology* (1830) from Captain Fitzroy. He began to devour the ideas quite avidly, and quickly came to recognize the superiority of Lyell's uniformitarian methodology over previous theoretical approaches. He became eager to test Lyell's ideas at the first landfall in the Canary Islands.

The Beagle set sail in late December and did not return until nearly five years later in 1836. During the years 1831-1836 Charles tested Humboldt's conclusions and Lyell's *Principles* very successfully, first around St. Jago in the Canary Islands and, later, more fully in Brazil, Uruguay, Argentina and Chile. He sent back frequent letters detailing his discoveries, ideas and conclusions (Burkhardt

& Smith, 1985). Boxes of specimens and new conjectures, e.g. concerning the origins of large extinct fossils like the *Megatherium*, were frequently received by Henslow and read out both in Cambridge and at the Geological Society of London (GSL). Because of the episodic and tardy arrival of the ship's post, Charles had no idea of the scientific excitement that his labours were causing in the British Isles. Indeed at times, he was in despair at the lack of replies to his many queries and requests for guidance. Not for the first time did he have self-doubts about his future as a geologist and naturalist. Charles collected data and generated ideas avidly and eventually published them in his *Journal of Researches into geology and natural history...* (1839), *Geological Observations on South America* (1846) (including a vivid account of the effects of the Conception earthquake and a prescient theory of magmatic differentiation beneath the Andes).

As the survey moved into the Pacific, he had formulated his own theories on coral islands before he had landed on one (published seven years later in 1842) to counter those of Lyell and others. He added many facets to his notes on volcanic phenomena that were eventually published as *Geological Observations on Volcanic Islands* (1844). All this, and much more, served to confirm his standing as the most experienced and promising of the young geologists and he was even more lionised back at the GSL before the Beagle docked in Falmouth in 1836.

2.6. A return to a scientific life in London

After 5 years travelling, and experiencing many events (including persistent sea sickness), which he had no wish ever to repeat, he hurried home to the Mount in Shrewsbury. He was soon recounting his adventures to his father and sisters and then to his uncle Jos and his cousins at Maer Hall. Soon he settled in London in order to unwrap his voluminous specimens and enlist the support of experts in examining them and writing up the implications. At first he resisted, then accepted, the post of Honorary Secretary to the GSL (1838-1841). This act placed him at the very centre of scientific influence and power in London. Amid the radical political ambience of the 1830s and 1840s (strikes, Chartist riots, pamphleteering re slavery, sectarianism), he settled down to write and edit the many books and reports stemming from the voyage. Alas! he soon suffered from bouts of illness which have never been adequately accounted for, and these restricted his further development as a geologist. His only forays relating to field geology after his return to England were to describe and account for: (i) the Parallel Roads of Glen Roy in terms of marine erosion and uplift comparable to that of Chile (an interpretation which caused him acute embarrassment in future years) and (ii) the landscape of North Wales in terms of glacial erosional and depositional processes (as introduced into Britain by Louis Agassiz and championed by Rev. William Buckland, the latter being readily converted to the new land-ice faith).

During all this time, Charles returned as frequently as possible to Shrewsbury and Maer, often arriving exhausted by stagecoach at the Lion Inn in Shrewsbury or the Crown Inn, Stone, Staffs, at which latter place he was picked up by phaeton for transfer to Maer. While at Maer, he continued his exploration of the delightful parkland landscapes fashioned by Jos. Wedgwood II and Johnathan

Webb, the landscape architect, between 1805 and 1822. Miraculously this landscape has survived intact to the present day and is preserved as a Grade I landscape (hence on a par with Castle Howard and Blenheim), together with the grade II Maer Hall and grade II tunnelled approach way to the hall and village. His long walks often took him further afield into the local countryside and to the estates of the gentry, for example at Keele, Betley, Trentham and Swynnerton.

2.7. The origin of soil

Very soon (1836-37) Charles heard from Uncle Jos of the strange dispositions of the soil in a field alongside William Dabb's peaty bog meadow and Croft. "Burnt marl, lime and cinders", and other recognizable material, spread on the top of the field 80 years ago to counteract poor drainage and acidity, was now found 12-14 inches below the surface. Uncle Jos offered the conjecture that this was somehow due to the actions of the worms, which were so common. This idea became known in the family as "Ye Maer Hypothesis" of Josiah II (Elizabeth Wedgwood to Charles Darwin 10.11.1837, in Burkhardt & Smith, 1985, p. 55). At that time there were contrasting theories of the origin of soil, the Wernerian view being that soil was the last deposit of a retreating universal ocean. Elie de Beaumont (1829) had followed this by opining that soil was the most unchangeable, inert material on Earth. By contrast Hutton and Lyell explained that soil was due to the underlying rocks being physically and chemically broken down, so providing a mantle of material which could be eroded, transported, deposited, buried, lithified, uplifted and recycled endlessly. In both theories, the role of earthworms was not particularly considered. Charles seized upon Jos' conjecture and determined to put it to observation and investigation, thus beginning an interest in experiments on worms which went on and off for the whole of his life. Charles produced his initial findings in a paper to the GSL, which was very well received in 1837 (Darwin, 1838, 18?40). These observations caused quite a stir, not least because his calculations of the annual turnover of soil by passage through the alimentary systems of worms amounted to 18-tons/acre/per annum. Charles went on to take an interest in this subject all his life, culminating in his experiments at Downe in Kent, involving the construction of his intriguing apparatus on the lawn at Down (sic) House and his best-selling book *On vegetable mould....* (1881).

By 1838, aged 29, Charles contemplated life without marriage and decided that he ought to write down the pros and cons of seeking a wife. Having answered himself in the affirmative, he thought that he should first ask his cousin Emma Wedgwood of Maer Hall whom he had known from childhood and who had supported him so staunchly whilst he was away at sea. On Sunday 11th November 1838, he summoned up courage to propose to her and, somewhat surprised and taken aback, she accepted hurriedly before dashing off to teach the children of Maer Sunday School. The couple were married on 29th January 1839 in St Peter's Church, Maer, in front of a very small congregation of family friends. The registration of the marriage is the second entry in what was, until very recently, the current register (!). A copy of the now famous certificate is conveniently posted on the notice board in the vestibule of the church. Students of all ages today are always somewhat abashed to realise that Emma, then aged 31, was to bear Charles 10 children in subsequent years.

2.8. The discovery of dolerite dykes in North Staffordshire

Charles continued to visit Maer frequently and his interest in the area became both professional and social. In London, he was in monthly contact with Roderick Impey Murchison (1792-1871) (a rising star of the geological firmament who was in process of composing his massive tome "the Silurian System" eventually published in 1839). Murchison had been scouring the Welsh Borderland looking at the Lower Palaeozoic strata and their cover rocks. He had observed the intrusive and extrusive rocks of the Breidden Hills (now known to be Ordovician) and the trend of the landscape features to which the geological structures give rise. He recognized two lines of hill masses in the Breiddens. He noted that they were aligned northeast towards Clive, Grinshill and Acton Reynald (where dyke rocks cutting the New Red Sandstone had been discovered by the local architect John Carline II and pointed out to Murchison on two visits in 1834). He believed that the two lines of hill masses including the dykes, passed onwards towards Hawkstone and eventually passed through Goldstone Common. The trend was said to be continued in the high ground in Staffordshire in the Ashley Hills and beyond. Murchison was keen, perhaps over keen, to relate these phenomena to the latest and most beguiling theory of the structural origin of mountains; the theory of "Lines of Elevation" and "Craters of Elevation" of Elie de Beaumont (1829 and subsequently) and Leopold von Buch respectively. Confidently he predicted that dykes would be found associated with the high ground of North Staffordshire. As was his style, Murchison dashed off a letter explaining his ideas to his friend Robert Garner, physician-surgeon at the Infirmary in Stoke-on-Trent and personal doctor to Josiah Wedgwood II (Kirkby, 1894; Anon, 1986). Garner did not find any dykes personally, but Kirkby notes (1894, p.130) that "a dyke was discovered some time afterwards by Darwin who was on a visit to Josiah Wedgwood at Maer and told him of it. Mr Wedgwood told Mr Garner, and the latter has it marked in (sic) the map illustrating his History of Staffordshire" (1844, p.206 and map). The hand-coloured map of 1844 shows a dyke extending northwest-southeast for two km from a point 0.2 km north of the Trentham-Stableford Road south-eastwards to an area southeast of Harley Farm, this extent being no doubt an exaggeration due to the difficulty of locating and depicting a narrow dyke on a small-scale map. Garner's account by contrast (1844, p.206), only states that Josiah Wedgwood pointed out the dyke to him "by the road from Hanchurch to Stableford" where modern studies have confirmed the presence of two or three dykes (Thompson and Winchester, 1996). Darwin is furthermore credited with discovering a second dolerite dyke (now known to be of the same swarm and suite) cutting across a narrow lane near Butterton Lodge Gates (ibid.). The exact date of Darwin's discoveries of the dykes is not known but consideration of all the circumstantial evidence suggests that this occurred in 1842.

2.9. The origin of species

The year 1842 was the year in which Charles first attempted to commit his long-lasting thoughts on the transmutation of species to written format. He had begun keeping notebooks on this topic 5 years previously. The context in which such ideas gestated owes much to his geological background. During his years at sea and afterwards, he had tested the ideas of Hutton, Lamark, Cuvier and Lyell and was in the process of slowly developing his own. Quite crucially, he had witnessed the slow growth of the Huttonian concept that we now know as "Deep Time". By 1840, a more or less reliable geological timescale had been developed through the efforts of William Smith and others and was summarized by John Phillips in the *Penny Cyclopaedia* around 1840; thus, aeons of time appeared to be available for the origin and development of life and landscapes.

Darwin had pondered also on the multiplicity of species, their geographic distribution, their differences, their competition for food and survival etc. and he had eventually read Malthus' essay in 1838. Thus despite episodic bouts of illness, he was well enough to sit down at Maer Hall at the age of 34 during May 1842 to write a "garbled" pencil sketch of only 35 pages of "crabbed elliptical scrawl" (Browne, 1995) which nevertheless epitomized his salient ideas on the transmutation of species over geological time. This initial outline of what became known as Natural Selection was revised in a longer draft of c. 250 pp. at his new home at Downe in Kent in 1844. The final version was projected to be a three volume work, but this never got off the ground in the light of the arrival of Alfred Wallace's rival paper, disclosing similar ideas, which was first received in London in 1856. Eventually *Origin of Species* was published as one volume in 1859.

There have been endless speculations about the non-publication of such a major scientific and philosophical work in the years between 1844 and 1859. These relate to Darwin recognising several factors: the misgivings of his much-loved and deeply religious wife; the likelihood of his suffering social ostracism as a consequence of publishing heretical works so counter to the teachings of Genesis and/or Natural Theology, and the likelihood of public criticism bringing on further bouts of (psychosomatic?) illness.

2.10. A quiet but scientifically productive life at Downe, Kent, 1842-1882

The rest of Darwin's career after he "retired" to Downe in 1842, to work quietly on largely non-geological topics, need not concern us. Suffice to say that it was nearly 20 years after he returned from the voyage of the Beagle before his biological publications began to outnumber his geological contributions to our scientific and philosophical heritage. Students may be encouraged to ponder why Darwin at the height of his powers as a scientist in the 1860s was never ennobled or knighted (as had been Henry de la Beche, Lyell and Murchison before him), and why, after this rejection, he came to be buried close to his hero Lyell in Westminster Abbey.

The house, gardens, greenhouse-laboratories and sand walk at Down (sic) House are now managed by English Heritage. Having been renovated and conserved at great cost, they were opened to the public in April 1998. They are well worth a carefully planned school visit.

3. SAMPLE ITINERARIES FOR ALOCAL OR NATIONAL HEROES@ FIELD TRIPS

The following are itineraries for half-day or full-day field visits in North Shropshire (the Shrewsbury area - Excursion A) or to Maer, the Maer Hills and adjacent locations of Darwinian interest (Excursion B) which seek to capitalise upon the educational opportunities which are offered by these areas. For the locations on these field trips, please consult the original figures and maps referenced in the original article (Thompson, 1999).

Safety. Almost all the localities visited present few (other than normal) dangers to visitors unless a visit is made to the sand and gravel pit where hard hats and attention to water hazards, the movements of lorries, dumper-grader trucks and overhead machinery will be needed.

Excursion A. The Shrewsbury area of North Shropshire (Figure 4 in the original article).

1. St John's Hill, the street in which Charles Darwin's parents, Dr. Robert and Susannah Darwin, first made a home.

2. The Crescent, the lane to which the family soon moved.

3. The Mount, the large house that was erected by Dr. Robert Darwin. This was Charles' birthplace (1809) where he and his brother Erasmus carried out their chemical and gardening experiments. This is the place where Adam Sedgwick dismounted, hitched his horse and stayed the night before embarking on the joint field excursion with Charles in North Wales in August 1831.

4. The new St. Chad's Church, where Charles was baptized at a red sandstone font rescued from the Old St. Chad's church (not the present crinoidal Carboniferous limestone structure).

5. The former preparatory school (No. 13 Claremont Hill), which Charles attended, and from whose back window he witnessed the ceremonial burial of a Dragoon soldier - an event which had a profound effect on a small boy so soon after his mother's death in 1817.

6. The former free grammar school of Shrewsbury, established for the sons of the town's rich and powerful traders. This building, now housing the town's library, is fronted by the blackish larvikite (syenitic) statue of Charles Darwin, the school's notably unsuccessful but most famous student.

7. The Morris' Hall, outside which lies **The Bellstone**, a so-called "granite" boulder (actually of ignimbrite and pyroclastic-flow origins; of Ordovician age and matching rocks known in outcrop in the Lake District of northwest England). In the absence of land-ice glacial theory (until Louis Agassiz's visit in the early 1840s), the origins of the boulder could not be explained, except by reference to the Flood or to floating icebergs.

8. Preston Montford Church and Churchyard, the place where Dr. Robert Wareing Darwin and Susannah (nee Wedgwood), Charles' father and mother, are buried in a tomb on the south side of the church. The route to this place illustrates the type of open countryside through which Dr. Robert's coachman would drive his yellow chaise when the Doctor was on his rounds to see patients and/or arrange financial loans.

9. Woodhouse, the country home of William Mostyn Owen to which Charles Darwin made frequent family visits. It was here that he enjoyed his earliest attempts at wooing girls, notably Fanny Owen.

Excursion B. The Maer and Maer Hills area in North Staffordshire (Figures 3 and 5 in the original article).

1. Travel to **Maer village** in order to set the scene, passing **Whitmore Station** on the former Grand Junction Railway line (opened 1837). It was from this place that Charles and Emma Darwin departed on honeymoon to London on January 29th, 1839.

2. Travel into the Maer hills via Slymansdale to **the entrance to Camp Hill House and the small quarry**. Investigations hereabouts can include the identification of rock types, a small normal fault, the rock succession and the location of three types of ecology and soils on the Keele Sandstone/Mudstone (290 Ma (millions of years old) (Carboniferous) and the Kidderminster Pebble Beds (245 Ma, Triassic).

3. Travel east to west along the foot of outcrop of the Kidderminster Pebble Beds identifying the escarpment and its dip slope to the south. It was in this area of the Maer Hills and the Pebble Beds that a planning application for a large sand and gravel pit was refused and a scheme for a single golf course was approved but not carried through. Discuss the criteria for accepting/rejecting such an application.

4. Observe that **the escarpment east-west ends abruptly just past Holloway Lane Farm** and is replaced by an escarpment aligned north-south due to the action of differential erosion along the Sidway fault line.

5. Travel to **Willoughbridge and the working sand and gravel pit of Hanson Aggregates** (formerly the Amey Roadstone Company (ARC)) in order to reconstruct the stages in the establishment of such a pit and the environmental effects of its presence. This allows an understanding of the consequences of erecting such a pit in the middle of the Maer Hills around grid square should a planning application be granted. Details of a dozen or so tasks for students to carry out in the gravel pit will be found in Thompson (1999).

6. Travel onwards to the Bogs, a peat-filled glacial overflow channel, and locate **William Dabbs' cottage.** The surrounding croft (meadow) is the place where Josiah Wedgwood II interested Charles Darwin in devising their theory of the nature of origin of soil as a result of the actions of earthworms (Darwin, 1838, 1840?).

7. Travel back to Maer village and turn westwards up the line towards Bates Farm. **At a right angle** in the lane, gain **a vantage point on the old Woore/Maer Road** (now a grassy track way). Cross two stiles and in 20 m and gain a view over the Maer Hills to the northwest. Evaluate the geological structure of the area, the effects of the Madeley and Sidway Faults in relation to the Maer Hills structural block (Figure 3 in Thompson, 1999). Evaluate the landscape and test whether the development of a sand and gravel pit of the Willoughbridge kind could ever be hidden from view in the Maer Forest or whether it would stand out as an unacceptable blot on the landscape as conservation groups contend.

8. Walk back down the lane southwards for 300 m and savour the views of **Maer Hall** to the east which is a grade I historic landscape comparable with those of Blenheim in the Cotswolds and Castle Howard in Yorkshire. Evaluate the wisdom of permitting a planning application for an 18-hole golf course with the 17th fairway and the 17th and 18th greens within sight of the hall. This is the hall where the first pencil sketch of the Origin of Species was written in the summer of 1842, five years after Darwin opened his transmutation notebooks. This is the sublime environment in which Charles courted his cousin Emma, wrote down the pros and cons of marriage and plucked up courage to propose on November 11[th.] 1838.

9. Walk to **St. Peter's Church, Maer** (Figure 5 in Thompson, 1999) where the marriage of Charles and Emma took place. Locate the copy of the entry in the Births, Marriages and Deaths Register, relating to their marriage on January 29th 1839, is posted on the notice board in the vestibule. Outside the church, note the position and orientation of the grave of Josiah Wedgwood II and his wife Elizabeth and the significance of the materials of the gravestone.

10. Travel from Maer following the route of one of Darwin's long walks or horse rides via Chapel Chorlton and Stableford towards Hanchurch. **Stop in the Hanchurch Hills** where Darwin is reputed to have located a dolerite dyke under or alongside the road from Trentham to Stableford around 1842. He had been primed to do this by using a woefully inadequate theory of Roderick Impey Murchison. Reflect on the great value of using totally ill founded theories in science. Travel onwards to **near Butterton Lodge gates**. Here, on the former main road from Market to Newcastle, Darwin is credited with locating a further exposure of a Tertiary Dolerite/Basalt dyke using the same ill-founded theory.

REFERENCES

Anon, (1603). *The authorised version of the Bible*. London.

Cuvier, G. (1813). *An essay on the theory of the Earth*. Edinburgh.

Darwin, C.R. (1831). (A letter written at Maer to Dr R.W. Darwin, his father, dated 31st August 1831, stating Charles Darwin's opinions for and against an acceptance of the offer to travel as a naturalist with Capt. Fitzroy on HMS Beagle). Cambridge University Library DAR 223 & DAR 97 (Ser. 2): 10.

Darwin, R. W. (1831a,b). (Letters written on 30th August and 1st September 1831 from Shrewsbury to Josiah Wedgwood II, the first strongly objecting to his son Charles' plan to travel as a gentleman naturalist with Capt. Fitzroy, the second withdrawing all objections to the idea). Keele University Library (Wedgwood/Moseley 96).

Darwin, C. R. (1839). *Journal of researches into the geology and natural history of the various countries visited by H.M.S. Beagle*. London: Henry Colburn.

Darwin, C. R. (1842). The Structure and Distribution of Coral Reefs. Part 1 of *The geology of the voyage of the Beagle*. London: Smith Elder.

Darwin, C. R. (1846). Geological Observations on South America. Part 3 of *The geology of the voyage of the Beagle*. London: Smith Elder.

Darwin, C. R. (1859). *On the origin of species by means of natural selection, or the preservation of favoured races in the struggle for life (First edition)*. London: Murray.

Darwin, C. R. (1862). *The various contrivances by which British and foreign orchids are fertilised by insects*. London: Murray.

Darwin, C. R. (1868). *The variation of animals and plants under domestication*. London: Murray.

Darwin, C. R. (1876). *The effects of cross and self-fertilisation in the vegetable kingdom*. London: Murray.

Darwin, E. (1789). The loves of the plants, a poem with philosophical notes. Part 2 of the *Botanic Garden*. Lichfield.

Darwin, E. (1791). The economy of vegetation. Part 1 of the *Botanic Garden*. London.

Herbert, S. (1993). (Letter dated 25.3.1993 from one of the world's foremost Darwin scholars requesting the conservation of the Wedgwood Parkland Estate and the Hall in the interests of the history of science). Baltimore USA, The University of Maryland.

Hutton, J. (1785). Theory of the Earth. An Investigation of the Laws Observable in the Composition, Dissolution, and Restoration of Land upon the Globe. *Transactions of the Royal Society of Edinburgh*.

Hutton, J. (1795). *Theory of the Earth, with proofs and illustrations* (2 volumes). Edinburgh.

Hutton, J. (edited by Geikie, A.) (1899). *Theory of the Earth* (3rd volume). London: The Geological Society.

Playfair, J. (1802). *Illustrations of the Huttonian theory of the Earth*. Edinburgh.

Thompson, D. (1999). Charles Darwin's presence in North Staffordshire C. 1815-1842: The rationale for some educational initiatives. *Teaching Earth Science*, 24, 2, 52-68.

Wedgwood, Josiah II, (1831). (Letter written at Maer on 31st August 1831 to Dr Robert Wareing Darwin of the Mount, Shrewsbury, concerning the eminent suitability of Charles Darwin to join Capt. Robert

Fitzroy as a gentleman companion and naturalist on HMS Beagle). Cambridge, Cambridge University Library DAR 97 (Ser. 2) 6-8.

Whewell, W. (1837). *History of the inductive sciences from the earliest times to the present.* London: Parker.

Secondary Sources

Bowler, P. J., (1990). *Charles Darwin. The man and his influence.* Oxford: Blackwell, 250 pp.

Browne, J., (1995) *Charles Darwin; Volume 1 Voyaging.* London: Cape, 605 pp.

Burkhardt, F. and Smith, S. (Eds.), (1985). *The correspondence of Charles Darwin. Vol. 1 1821-1836.* Cambridge: Cambridge University Press, 702 pp.

Darwin, C. R., (1838). On the Formation of Mould. *Proceedings of the Geological Society of London II (no.52) for 1837-8*, pp. 574-576.

CHAPTER 8: A STUDENT CONDUCTED EARTH SYSTEMS FIELD INVESTIGATION

Hiroshi Shimono and Masakazu Goto, National Institute for Educational Policy Research, JAPAN

1. INTRODUCTION

Students in Japanese upper secondary schools are voicing distaste for science. In addition, few children elect to participate in field trips and other experiences in nature. School officials, teachers and parents view these problems with a great deal of concern. Global Science Literacy, with its focus on understanding our natural habitat, can be a program to help to alleviate both of these problems. Central to GSL, however, is providing students with positive experiences with nature. With limited school time and schools often located in crowded urban areas, how are such experiences to be accomplished?

Students in Japan have enough time to conduct a research project during the 40-day summer vacation normally scheduled from late July to the end of August. In some schools, science teachers assign students an inquiry-based study on a topic of their interest. Thus, some field experiences in GSL courses can be provided through students' inquiry-based tasks assigned in the summer vacation. Such experiences can foster improved scientific literacy needed to form an environmentally conscious society essential for Japan as it enters the 21st century. The Monbu-Kagaku-sho (formerly Monbusho) recommends that this type of student inquiry-based study should a part of the New Course of Curricula implemented in 2002.

2. A STUDENT CONDUCTED PROJECT

Here we introduce an example of a student's inquiry-based research, conducted for her summer vacation assignment, based on topics from the school curriculum. The student, Natsuko, was enrolled in the 9th grade at the Hongo lower secondary school. She was interested in the differences in temperature in different parts of the neighborhood in which she and her parents lived. She had learned the basic concepts in meteorology in 8th grade science lessons the year before. Therefore, for her summer vacation assignment, she investigated the distribution of temperature in several urban environments having different vegetation, topography and development. She conducted her research under her science teacher's advice assisted by her father. The science teacher advised and assisted her in deciding her topic and how to conduct her research so to complete it by the end of summer vacation. In her

V.J. Mayer (ed.), Global Science Literacy, 129–135.

research, she wanted to find out whether it was really cooler on the hill near her home, then on the river flood plain below. As a result of analyzing and summarizing her data, she found several patterns of temperature distribution related to the altitude of the measurement station. Her research was awarded the prestigious Prime Minister's prize in the student science research competition conducted by the Yomiuri Newspaper Corporation.

3. NATSUKO'S RESEARCH PROJECT

It is very warm and humid in Japan in the summer. Natusko lives in a suburban area in Yokohama City that is located at 35 degrees north latitude and 140 degrees east longitude. Her home is in a flood plain and surrounded by hills to the north and south. There are also some small valleys at the edge of the hills. Yokohama City has an annual average temperature of 20°C with the lowest temperature of - 3° C in winter and the highest one of 35°C in summer.

One afternoon in summer, while Natsuko was in the eighth grade, she noticed that it was cooler on the hill than in the flood plain near her house. Following that summer, she studied climate and weather of Japan in her eighth grade school science curriculum. Owing to her understanding the basic meteorology, she became more interested in the cooler environments on the hill and the differences in temperature between other nearby places. Therefore, she decided to investigate the temperature of locations near her home having different topographic features, vegetation and development for her ninth grade summer vacation assignment. Along with her teacher and father, she developed the following questions to guide her research:

• Is it always cooler on the hill than in the plain?
• Is there any difference in temperature between on the hill and in the plain?
• What causes the difference in temperature between on the hill and in the plain between summer and winter?

3.1. Procedures used by Natsuko

Natsuko decided to complete several tasks to help answer her questions:
• To collect data on the temperatures at certain time intervals from several locations (stations) near her home starting in winter and continuing on through the summer.
• To locate her stations on a topographic map and describe their geographical characteristics of elevation, vegetation and development.
• To investigate the relation between the geographical characteristics and the distribution of temperature with time of day and year.

Nastuko designated 53 measurement stations in the 20 square kilometers surrounding her home. There were 18 stations on the northern hill, 13 on the southern hill, 14 in the valleys and 18 in the flood plain. Each was located on her topographic map. There were many more houses and factories on the northern hill than on the southern hill. She made use of her family car for collecting her data because of the area that her research covered. The driving distance was about 30 km. She needed to collect the data at her stations at such special times as around noon, at nine p.m., and just before dawn the next morning. She designed her thermometer with a shield around its sensor so that it would not be influenced by air movement or exposure to direct sunlight. Her family drove their car for her and she measured temperature at each station by placing the thermometer out of the car window. She recorded the temperature, the time and the place using a tape recorder. It took her between 31 minutes in the early morning and an hour and 18 minutes in the afternoon to collect her data from all of her stations. The average time to collect data was 40 minutes in the early morning, one hour in the afternoon, and 50 minutes at night. She also recorded a description of the weather each day she took measurements, observing precipitation and cloud cover. Later, she transcribed these data on paper. She collected her data during at least one day per month for seven months.

As she collected and recorded her data, she tried to observe patterns in it. She thought about the relation between the geography and the temperature at each station by relating the station temperatures to the location of the stations on the topographic map. She also constructed cross sections of topography and temperature to assist her in thinking about possible relationships. Through performing these analyses over the seven months of her research, she reached the conclusion that there were three typical patterns of the relationship between the distribution of temperature and geography.

3.2. Natsuko's data

Figure 1 shows the aerial distribution of her temperature measurements. After analyzing her data, she found the following patterns in the temperature distribution on the hill, in the flood plain and in the valley.

On the northern hill the temperature:
• was higher than on the southern hill;
• was higher in the plain early in the morning in the early spring and in the winter;
• fell below 0 degree centigrade during the winter.

On the southern hill the temperature:
• was higher early in the morning in the early spring than it was in the plain;
• sometimes went down below zero degrees centigrade in the early spring and in the winter.

In the plain the temperature:

- was the highest except for the early morning in winter and in early spring;
- sometimes went down below zero degrees centigrade in the early morning in the winter.

On the low hill the temperature:
- was the second highest except for in the morning in the winter and the early spring;
- went down below zero degrees centigrade in the early morning in the winter.

In the valley bottom the temperature:
- was the lowest in the early morning;
- sometimes rose in the daytime;
- went down below zero degrees centigrade in the early morning in the winter.

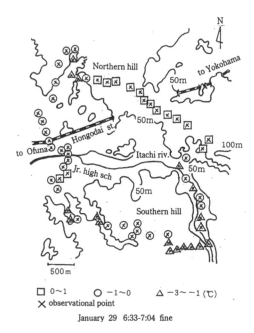

Figure 1. Aerial distribution of temperatures

3.3. Natsuko's analysis and conclusions

Natsuko analyzed and summarized these data in a series of figures and tables. From these, she concluded that there were three patterns of change in temperature (See figure 2). She classified 23 data points from the total number of 42 into these three patterns:

- Pattern A: The temperatures on the northern and southern hills were higher than those in the valley and at the same time, the temperature on the northern hill was higher than that on the southern hill. Pattern A was apt to appear on a fine day and under high atmospheric pressure. She concluded that the difference in temperature on the hills was likely due to heat emitted from the buildings in the housing area since the northern hill was more highly developed.
- Pattern B: The temperature on the northern and southern hills was lower than that in the valley and the plain where Hongo Junior High School is located. Pattern B appeared mostly in the afternoon. Area residents had noted that they felt cooler on the hills during the summer.
- Pattern C: The temperature on the northern hill was higher than that in the valley. The temperature on the southern hill was the same as or a little lower than that in the valley. People feel cooler on the southern hill than in the plain. She guessed that the northern hill is influenced more strongly by the artificial heat from development.

She found that seven A patterns and two B patterns occurred in the morning, seven B patterns and one C pattern in the afternoon and three B patterns, two C patterns and one A pattern at night.

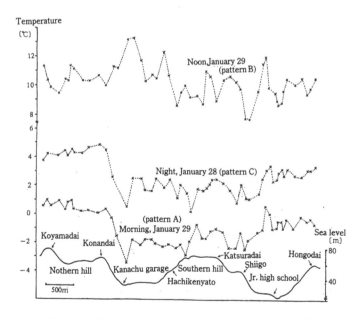

Figure 2. Temperatures associated with topography

3.4. Her opinions and feelings after completing her research

Natsuko expressed the following opinions after completing her study:
- "It is hard to get up early in the morning to measure the temperature."
- "I observed that the sound of insects, people's movement, and the smell of the air were different in the morning, in the daytime and at night."
- "I had thought it was always cool on the hill, but I had a lot of questions while I continued to do research and made the tables and graphs during my investigation."
- "I regretted a little that the climate on the hill was influenced by human beings. I think that each person influences our environment through emitting the heat energy from his house."
- "I was surprised that I could learn a lot about the environment, using only one thermometer."

4. SUGGESTIONS AND RECOMMENDATIONS

As you learn from Natsuko's research, students can conduct good research on their local environment without special instruments. Starting in 2002, the *New Course of Study* will be introduced in Japan. In the *New Secondary School Curriculum*, this kind of student inquiry-based research will be recommended. Science research related to natural environments will be especially useful since the *New Course of Study* also endorses environmental education. Students will learn through this kind of research of the effects of urbanization so prevalent in Japan. Science teachers must be prepared to advise, support and guide students in their inquiry-based research, and to foster the scientific view and way of thinking.

Several ministries in cooperation with the former Monbusho improved their support for students' inquiry-based study. Monbusho conducted *The Children's Plan All Over Japan* from 1999 to 2001. This program promoted various environmental education activities in each district as they move to the five-day school system:
- The Broadcast for Children through Satellite Transmission - Children can watch a sports figure, a first-class scientist or technician, and leaders of environmental issues through televised appearances in a local public hall, library and museum.
- The Village for a Long-term Experience in Nature (in cooperation with The Agriculture, Forestry and Fisheries Ministry) - Children can have a two-week nature experience, agricultural experience, or environmental learning during the summer vacation in facilities like an agricultural house or youth hostel.
- The Children's Park Ranger (in cooperation with the Environment Agency) - Children, appointed as park rangers, have an opportunity to participate in conservation and nature activities with the assistance of park rangers and volunteers in a national park.
- The Promotion of Children Doing Science and Making Some Work (in

cooperation with the Science and Technology Agency) - Children can have hands-on activities in museums and schools on Saturday.

- The Opening of Universities and Institutes - Children can have opportunities to learn various subjects from specialists on Saturday in universities and institutes.

- The Arrangement of the Experiential and Playing Field for Children (In cooperation with the Construction Ministry and Agricultural, Forestry and Fisheries Ministry, and the Marine Agency) - Children can have experimental activities on a river, in a fishing port, or local nature supervised by educators, leaders of youth associations, and local citizens.

Such activities for environmental learning will be arranged much more frequently in the future by the various ministries. Students will be able to make use of as many opportunities as possible, like these, for inquiry-based study during the summer vacation

CHAPTER 9: USING HISTORICAL EVENTS TO DEVELOP ETHICAL AND AESTHETIC ATTITUDES

Fernando Lillo, Santiago de Compostela University, SPAIN
José Lillo, Vigo University, SPAIN

1. INTRODUCTION

We are in a transition during which the globalization of politics and economies has also influenced the need to include a concern for improved global education in pre-college curricula (Yus, 1996, 1997). At present, correcting the lack of an holistic view in school education in some countries such as in Spain is being attempted through employing interdisciplinary curricula. In this sense, one of the authors (Lillo, F., 1999) has developed a proposal on behalf of the Ministry of Education for training Secondary Education teachers. The proposal suggests a cross-curriculum approach among natural and social sciences and classical culture. Taking this interdisciplinary proposal as a basis in concordance with the GSL objectives, we have developed sequences of activities following the ways that Greeks and Romans understood nature. The objectives for these activities are to develop positive ethical and aesthetic attitudes about the role of human beings in nature. They cover conceptual, procedural and attitudinal content contained in Spanish science curricula through the analysis of classical texts, resolution of questions about them and debates on environmental impacts introduced by humans throughout history. They can be adapted easily in other contexts.

2. RELATIONSHIP OF THE ACTIVITIES TO GSL

GSL uses the concept "Earth system science" developed by the Earth System Science Committee (ESSE, 1988) to be appropriate for coordinating international investigations on systems that operate in and on the Earth. Through this research effort, programs have been developed that offer "on-line" access to up-to-date research data, some of which can be used in science lessons. The curriculum concept of "Earth Systems Education" stemmed from that activity. Subsequently social, ethical and aesthetic aspects have been incorporated as well as concepts of cultural diversity, globalization and other social aspects to form the construct of Global Science Literacy. A central theme in GSL is the consequence of the use and abuse of the Earth's resources. These ideas have also been extended to teacher training consideration (Mayer, 1995).

In proposing the following activities, we have made reference to Earth Systems Education Understandings One and Two (refer to Table 1, Chapter One). These were developed initially for use during a series of teacher preparation and curriculum development programs conducted at The Ohio State University and the University of Northern Colorado. Subsequently they have been followed by secondary school teachers in developing Earth Systems Education curricula. The following activities have been put

V.J. Mayer (ed.), Global Science Literacy, 137–146.

into practice with Spanish high school and first university course students, and also in a recent workshop on training Galician science and mathematics teachers in its annual meeting, under the title "The view of ecology in the classical world: A proposal to develop some interdisciplinary contents and attitudinal objectives in secondary and post-compulsory education" (Proceedings of the XII Congress of ENCIGA. Science Teachers Association of Galicia, Spain, September 1999, pp. 131-145).

2.1. Activities regarding the Greeks' attitudes toward nature

2.1.1. Nature and gods
a. Natural things as sacred:

Originally Greek sensitivity saw all nature as sacred, on many occasions because of peoples' impotence before certain natural phenomena that overcame them and their structures. This caused them to associate gods with natural environments and to locate their sanctuaries in places of special beauty.

Activities: Read the following text carefully and then answer the questions that follow:

The attitudes of the early Greeks toward nature were shaped by their religion. They saw the natural environment as the sphere of activity of the gods. Greek religion was in large part the worship of nature, and the old Greek gods were essentially nature deities. The gods ruled nature, they appeared in it, they acted through it; therefore, human activities which affect the environment often were seen as involving the interest and reaction of the gods. Zeus was a weather god, wielding the dreaded thunderbolt. He made the wind blow favorably or disastrously, moved the clouds, stirred up storms, thundered, lightened, and roused the waves of the sea. (...) Poseidon, originally the underworld god of springs, the earthquake god who shook the mountains, became the preeminent god of the sea. Athenea's patronage of wisdom and defensive warfare were importations; her earlier concerns were the owl, the serpent, and the sacred olive tree (...) Artemis, far from being only the divine huntress, was from old "Artemis of the wild wood", the protectress of wild animals and the guardian of the wilderness, bearing the very ancient title Potnia Theron, "Mistress of Beasts". She was a fertility goddess who encouraged the multiplication of birds and animals. -- J. Donald Hughes, Ecology in Ancient Civilizations, pp. 48-49.

Questions:

1. What relationship does Greek religion have with Nature?
2. Did gods perform special actions in Nature? Describe the relationship that each god has with Nature. Keep in mind that, sometimes, it may not be the only characteristic of the god.

Greek God	Associated Natural Characteristics
Zeus	
Poseidon	
Athenea	
Artemis	

b. The sacred contributes to the conservation of the natural environments:

Activities: Read the following text carefully and answer the questions that follow:

The association between the gods and trees was particularly close in the Greek mind, and all Greek altars and places of worship were originally outdoors in groves of trees. (...) Probably the Greeks first worshiped in groves and only later built temples within them; it is certain that the association of the two was so deep that temples were always set in groves of trees, and when a grove was not available, as was the case on the rocky summit of the Acropolis when the Parthenon was built, holes were excavated in the solid rock and trees were planted in rows to flank the temple. The Greeks could not conceive of a sacred area without trees. The result of this attitude was the preservation of the sacred groves from fire, grazing, and the ax, so that they remained like parks in a semi-natural state, and the trees in them grew huge with age. -- J. Donald Hughes, *Ecology in Ancient Civilizations, p. 50.*

Questions:

  1. What was the result of the relationships between gods and trees?
  2. Do trees exist in modern religious sanctuaries? Think of Fatima (Portugal), Lourdes (France) or Covadonga (Spain).
  3. What places are chosen for the construction of natural sanctuaries? Give concrete examples from your personal experience.

2.1.2. Nature as an object of knowledge
One of the Greek attitudes toward nature was the desire to know and to understand it. This caused many philosophers to think carefully about nature and write down their ideas. Among them, Aristotle and his pupil Theophrastus were those who left the most information about nature in their writings. We would like to present an observation that could be qualified as ecological. Keeping in mind, however, that ecology as a science did not exist in Antiquity. We could make observations, however, that would now be

classified as ecological, because they had to do with relationships between different living beings.

The philosopher Plato refers us to the observations that the Greeks had made about the animals and the balanced relationships among them. Instead of boring us with theories he tells us a myth or story. In this myth the gods asked Prometheus and Epimetheus to grant qualities to the created animals.

Activities: Read the following text carefully and then answer the questions that follow:

Epimetheus said to Prometheus: "Let me distribute, and do you inspect." This was agreed, and Epimetheus made the distribution. There were some to whom he gave strength without swiftness, while he equipped the weaker with swiftness; some he armed, and others he left unarmed; and devised for the latter some other means of preservation, making some large, and having their size as a protection, and others small, whose nature was to fly in the air or burrow in the ground; this was to be their way of escape. Thus did he compensate them with the view of preventing any race from becoming extinct. And when he had provided against their destruction by one another, he contrived also a means of protecting them against the seasons of heaven; clothing them with close hair and thick skins sufficient to defend them against the winter cold and able to resist the summer heat, so that they might have a natural bed of their own when they wanted to rest; also he furnished them with hoofs and hair and hard and callous skins under their feet. Then he gave them varieties of food-herb of the soil to some, to others fruits of trees, and to others roots, and to some again he gave other animals as food. And some he made to have few young ones, while those who were their prey were very prolific; and in this manner the race was preserved. -- Plato, *Protagoras* 320d-321b.

Questions:

    1.   How does Epimetheus prevent any species from annihilation? Give examples of the compensatory characteristics that he gives to certain animals.

    2.   What escape abilities does Epimetheus give to the animals?

    3.   What remedies does Epimetheus grant the animals for protection from the rigors of the different seasons?

    4.   How do the different animals feed? Describe the food chain of the animals. Use an explanatory drawing.

2.1.3. The impact of the Greek world on natural environments
a. The effect of deforestation in Greece:

We are used to seeing the Greek landscape as a landscape without too many trees. Probably half of the surface of Greece was originally covered with trees. Now forest covers only about a tenth of the country. The deforestation was due to several needs of the population. When contemplating the Acropolis of Athens and its magnificent temples we do not usually observe the landscape of the denuded mountains of the region of Athens that surround it. The philosopher Plato, who lived in the IV century B. C., realized the serious consequences that deforestation brought to Athens. He expressed sorrow about the noxious effects of indiscriminate pruning upon the landscape.

Activities: Read the following text carefully and then answer the questions that follow:

What now remains compared with what then existed is like the skeleton of a sick man, all the fat and soft earth having wasted away, and only the bare framework of the land being left. But at the epoch the country was unimpaired, and for its mountains it had high arable hills, and in place of the "moorlands", as they are now called, it contained plains full of rich soil; and it had much forest-land in its mountains, of which there are visible signs even to this day; for there are some mountains which now have nothing but food for bees, but they had trees not very long ago, and the rafters from those felled there to roof the largest buildings are still sound. And besides, there were many lofty trees of cultivated species; and it produced boundless pasturage for flocks. Moreover, it was enriched by the yearly rains from Zeus, which were not lost to it, as now, by flowing from the bare land into the sea; but the soil it had was deep, and therein it received the water, storing it up in the retentive loamy soil; and by drawing off into the hollows from the heights the water that was there absorbed, it provided all the various districts with abundant supplies of spring waters and streams, whereof the shrines which still remain even now, at the spots where the fountains formerly existed, are signs which testify that our present description of the land is true. -- Plato, Critias 111 b-d.

Questions:

1. What consequences has the erosion of the "fat" and "soft" part of the landscape had?
2. How is the region described before deforestation?
3. Why didn't the land with trees and abundant soil lose the water? What are the benefits of a landscape with abundant springs?
4. Have you been witness to any similar change in a landscape familiar to you? Are the consequences the same as for Plato?

Activities: Read the following text carefully and then answer the questions that follow:

But the activities of lumbermen and charcoal burners were not the only forces destroying the Greek forests. Forest fires, many of them set on purpose by shepherds, raged unchecked. And natural regeneration of the forests, which is slow in the dry Mediterranean climate, was usually prevented by the practices of grazing. (...) Even ordinarily grass-eating animals like sheep and cattle are forced to eat leaves and twigs in the Greek hills, and goats, which browse on bushes and small trees by preference, will also climb into large trees and eat the foliage. Goats made permanent the deforestation of thousands of square miles of Mediterranean hillsides by eating every

.

seedling tree that ventured to show its head, until there were no more left. -- J. Donald
Hughes, *Ecology in Ancient Civilizations*, p. 69.

Questions:

1. What other factors, besides the cutting of trees, contributed to the deforestation
    of Greek landscapes?
2. Why do you think that shepherds set forests on fire? Nowadays we suspect that
    many fires are caused by arsonists. Do you know what or who could have
    caused them and why?
3. Do you believe we should wait for natural regeneration?
4. Do you believe that these deforestation agents still act today? Give some
    examples.

b. The extinction of some species:

The destruction of forests and hunting activities put an end to some species, although
the impact was smaller than that carried out by Roman civilization.

Activities: Read the following text carefully and then answer the questions that follow:

*Hunting was a sport for the nobility of Greece, whose exploits in pursuit of wild boar
and other creatures are celebrated in literature. More animals and birds were killed to
protect domestic animals and the growing crops; the Homeric simile of a lion pursued
by herdsmen is well known. Large numbers of smaller game were simply killed for
food.(...) The larger predators were decimated. The lion and leopard were extirpated
from Greece and coastal Asia Minor by the end of the Hellenistic Age. Wolves and
jackals were rarely seen outside the mountains.* -- J. Donald Hughes, *Ecology in
Ancient Civilizations*, p. 72.

Questions:

1. What were the reasons for hunting? What reasons survive at present? Do you
    believe that hunting actually contributes to the disappearance of wild animals?
    Give your reasons.
2. What profit do you think that Greeks could take from wild animals?
3. Do lions, leopards and wolves still exist in Greece?
4. What is the situation of wild animals in your own community?

2.2. Activities related to Romans' attitudes to nature

2.2.1. Nature as headquarters of gods
The most autochthonous Roman religion considered nature as endowed with religious
powers and the natural elements with supernatural powers. The primitive Roman gods
also controlled agricultural conditions.

Activities: Read the following text carefully and then answer the questions that follow:

Roman religion always had a strong sense of locality. Certain places seemed to the Romans to be endowed by nature with supernatural powers, or numina. (...) Not only groves, but individual trees, rocks, springs, lakes, and rivers were regarded in this way. The area of volcanic activity near Naples, with its sulfurous caverns, stem vents, hot springs, and shaking ground, was thought to be particularly numinous, and famous oracles were located there. Roman religion was not without reverence for wild places, forests, and mountains, but it was predominantly agricultural. For example, Jupiter was regarded as the bringer of rain. Anything that could be given a name seems to have had a deity in charge of it: Ceres was goddess of the growing grain. Liber, god of the wine, Robigus would protect the crops from diseases, and so forth. Beyond this, every major and minor activity of the farm had a deity who could be propitiated for its success, such as Vervactor for the first plowing, Repacator for the second plowing... -- Adaptation of J. Donald Hughes, *Ecology in Ancient Civilizations*, pp. 88-90.

Questions:

1. Do you know some other religion that attributes extraordinary powers to natural elements? Give examples.
2. What places did Roman religion perceive as sacred or having a special power? Do you know any place with a special or attributed power?
3. Is there in your town a special natural place that is related to something sacred? For example some fountain, rock or strange tree related to a legend or a saint.

2.2.2. The domain of nature
The desire to compete with nature and to dominate it is what Cicero expresses in the following fragment of *On the nature of gods.*

Activities: Read the following text carefully and then answer the questions that follow:

We have subjected into our domain even the quadrupeds, for transport, their speed and strength contribute to our own strength and speed. We get animals to carry loads, we impose our yoke on them. For our own benefit we use the elephant's sharp senses and the dog=s sagacity. We take iron from the bowels of the earth, necessary to plow the land; we find copper, silver and gold seams hidden in the depth of the earth, and we transform them into capable materials to use and to decorate. On the other hand, we use the wood resulting from cutting trees planted by man in their natural habitat, in part to warm ourselves, once lit, and to cook our food; in construction it is used to get rid of the excessive cold and heat. Wood also offers great possibilities to build ships, which are used to travel and bring us the necessary things for living. And, in spite of all

the violence nature, only we have the remedy against waves and wind, thanks to sea science. We also take advantage from sea products.
Similarly, we dominate everything that the earth offers, we benefit from the fields and the woods. We own the streams and lakes. We plant cereals and trees; we make the land fertile by conducting the waters. We separate, direct, deviate the river flow. In one word, with our own hands we dare to build a second nature out of the original. --
Cicero, *On the Nature of Gods*, 2.60 (152).

Questions:

1. Why did the Romans take advantage of their domination of the animals?
2. What did the Romans obtain from inside the Earth? How did they use them? Investigate the Romans' exploitation of gold in Las Médulas. (Explanation: Las Médulas is a nice Spanish place in which we can actually observe the *ruina montium* technique which uses water to obtain gold from detritic sediments.
3. What are the uses of wood? Do you believe that these uses contributed to deforestation?
4. What relationship did the Romans have with the sea? In spite of the triumphalism of the text, Greeks and Romans had a great respect for the sea and they didn't dominate it completely.
5. What part of the Earth did the Romans dominate? What does 'making lands fertile thanks to water conduction' refer to? Do you know any Roman aqueducts? Where are they?
6. Do you know a case in which Romans have diverted the course of the rivers? And a case of a similar thing happening in the present?
7. What is Cicero's conclusion?

2.2.3. The impact of the Roman World on nature
a. The extinction of some species:

As in Greece, the extinction of some bigger species was due to hunting and fishing. In Rome it is also necessary to keep in mind the use of wild animals in the circus and amphitheater shows that caused the extirpation of elephants, rhinoceroses and zebras from North Africa and the disappearance of the Lower Nile hippopotami.

Activities: Read the following text carefully and then answer the question that follows:

(Titus) after having dedicated an amphitheater and lifted some thermal baths next to it, he celebrated an astonishing gladiator show; he also offered a naval battle in the old naumachia, place where it also presented a gladiator show and exhibited five thousand different species of wild animals in one day. -- Suetonio, *The Life of Titus 7,3.*

Question:

What consequences did the species exhibited or killed in the Romans' games suffer?

b. The impact of Roman roads:

Activities: Read the following text carefully and then answer the questions that follow:

Beyond the walls of Rome in every direction stretched the Roman roads. They are one of the most significant ways in which the Romans left their mark upon the landscapes of Europe, Asia, and Africa. (...) They encouraged the development of agriculture, mining, and industry farther from the metropolitan center by providing access to more distant areas. Because of them, more forests could be felled, and plants and animals were transported and extended their ranges, through deliberate or accidental introduction. The roads increased human mobility and reduced the inaccessibility of more distant areas, both factors which amplified the impact on the natural environment of the Romans and those who followed them and continued to use the same roads. -- J. Donald Hughes, *Ecology in Ancient Civilizations*, pp. 124-125.

Questions:

1. In what ways did the Roman roads contribute to modifying the environment?
2. In what ways do current highways or high-speed trains contribute to modifying the environment?
3. Do you believe that road accessibility to certain places is a good thing? What risks does it bring? You can think of the example from some previously not very well-known tourist place to which access is facilitated.

c. Environment destruction as a war weapon:

Activities: Read the following text carefully and then answer the question that follows:

Destruction of human life and devastation of the natural environment went hand in hand in Roman warfare. The historian Tacitus has a British chieftain say or the Romans, "They make a desert and call it peace". Roman generals attempted to deprive their opponents or every means of subsistence and, therefore resistance. While they were often generous to those they conquered, they were capable of leveling enemy cities to the ground, as they did at Corinth. After destroying Carthage, the Romans ordered the fields sown in salt so that nothing would ever grow there again. -- J. Donald Hughes, *Ecology in Ancient Civilizations*, pp. 126-127.

Question:

How was environmental destruction used as a war weapon? What examples does the text give? Do you know of any recent examples?

3. FINAL EVALUATION OF ACTIVITIES

We were able to obtain several types of evaluations from individuals using these activities. They include recent evaluations given by our students; some obtained from some secondary school teachers who used the activities in their classes; and those received from the workshop participants in the Galician science and mathematic teachers annual congress. All of the evaluations show a positive attitude toward the activities and the use of this methodology, especially the interdisciplinary treatment of the classical world and the natural and social sciences. It has also been revealed as an appropriate way to develop critical thinking about the human role in nature throughout history and the necessity of considering ethical and aesthetic aspects which is not well covered in current science classes in Spain.

REFERENCES

Earth System Science Committee, (1988). *Earth System Science: a program for global change*, Washington: National Aeronautics and Space Administration.

Mayer,V. (1995). Using Earth System Science for integrating the science curriculum. *Science Education*, 79, 4, 375-391.

Yus, R. (1996). Temas transversales y educación global. *Aula de Innovación Educativa*, 51: 5-12.

Yus, R. (1997). *Hacia una educación global desde la transversalidad*. Madrid: Anaya-Alauda.

Ecology in Greece and Rome:

Cicerone (1967). *Sulla natura degli Dèi* (ed. de H. Pizzani). Milán: Arnoldo Mondadori editore.

Hughes, J. D. (1975). *Ecology in ancient civilizations*. Alburquerque: University of New Mexico Press. (Traducción española: *La ecología de las civilizaciones antiguas*. México: Fondo de Cultura Económica, 1980).

Lillo, F.(1999). *Cultura Clásica. En las áreas curriculares y en los Temas Transversales*. Madrid: Ministerio de Educación y Cultura-Narcea S.A. de ediciones, (See especially Unit: *Grecia y Roma ante la naturaleza* pp. 35-70).

Platón (1992). *Diálogos (Filebo, Timeo, Critias)*. Barcelona: Planeta-DeAgostini. Madrid: Gredos, 1992.

Platón (1993). *Protágoras-Gorgias*. Barcelona: Planeta-De Agostini. Mdrid: Gredos, 1993-4.

Suetonio (1996). *Vidas de los doce césares (De Nerón a Domiciano)*. Barcelona: Planeta-De Agostini. Madrid: Gredos, 1992.

CHAPTER 10: ASSESSMENT IN A GLOBAL SCIENCE LITERACY AND KOREAN CONTEXT

Jeonghee Nam, Pusan National University, KOREA
Victor J. Mayer, The Ohio State University, USA

1. INTRODUCTION

In this chapter, we discuss assessment in science education in a Korean context. We then identify the special characteristics of Global Science Literacy (GSL) and apply them to the role of assessment in future science education programs. In doing so, we define the concept of 'authentic assessment' in terms of GSL and discuss its significance in the real world of science curriculum development and implementation. Finally, we make judgments concerning the effect of Korean evaluation strategies on the likelihood of dramatically different curriculums, such as those based on GSL, being implemented in Korea, and by implication, other countries having evaluation systems similar to that of Korea.

Assessment includes practices designed to determine the success of a strategy or system of strategies used to produce change in a desired direction. In science education, this has usually meant the achievement and retention of scientific information and procedures by students, teachers or adult citizens. More recently, educators are using a new term 'authentic assessment'. It restricts the practical meaning of assessment in one sense and broadens it in another. It is more restrictive in the sense that any assessment should be performed in the same manner and context as instruction was carried out. It broadens what has been the practical or effective meaning of assessment in that it should measure the qualities intended to be created or enhanced by the instruction. Therefore the usual multiple choice or objective test of cognitive knowledge has limited although often valuable use in authentic assessment of science instruction. Other means must also be used, or used instead of the typical paper and pencil test.

Normally the acquisition of factual knowledge is only one, and at times perhaps a minor objective of science instruction. Also to be measured is a student's use of scientific problem solving strategies, creativity in arriving at solutions to problems, and certain laboratory or field skills. Also of interest is the student's attitude toward instruction and science itself. In instructional systems based on GSL, there are a number of aspects of learning that are unique and need to be monitored by assessment, especially those problem-solving methods referred to in Chapter Three as systems science methodology.

2. AUTHENTIC ASSESSMENT

Ideally, in performing authentic assessments or evaluations of instructional outcomes the vehicles and environments intended to produce those outcomes must

147

V.J. Mayer (ed.), Global Science Literacy, 147–156.

also be observed. This often means that the curriculum and teaching procedures need to be monitored to determine the student's opportunity to learn. Is the delivered curriculum consistent with the objectives stated for the outcomes of instruction? What instructional methods were or are being used? Do they have the potential to produce the intended learning outcomes? The student's ability to learn the intended outcomes must also be taken into consideration. What is the student's conceptual background? What experiences inside and outside of school has the student encountered relating to the intended outcome? Has the school environment been conducive to learning? Are there the necessary equipment and laboratory facilities? Is the physical layout of the classroom adequate?

In reality, therefore, authentic assessment can require a very complex system of measures and thus be very expensive and time consuming to implement. Seldom does the teacher, curriculum supervisor, school or system administrator have the resources to implement all of the necessary phases of a complete authentic assessment program. In reality, it normally comes down to the teacher and his or her immediate support staff, if there is any, to design and conduct assessments. The teacher and local school administer even those assessments that have been developed and required by higher educational authorities, such as those administered at the end of the lower secondary and upper secondary schools in Korea.

3. ASSESSMENT IN KOREA

3.1. General characteristics

Three features characterize educational assessment in Korea. The first one is the 'School Activities Record', which is designed to evaluate primary, middle and high school students based on their academic achievement and development of social behavior. The second feature is the *National Assessment of Educational Achievement*, which is to assess the quality of the entire educational system by providing information on overall educational achievement indicating where improvement might be needed. The third feature is the *College Scholastic Ability Test*, which is administered at the national level to identify eligible candidates for higher education.

The Presidential Commission for Educational Reform announced in 1995 that teachers should develop an individual student's *School Activities Record*. The purpose of this new evaluation system is to obtain diagnostic and formative information as well as summative information on the student's academic achievement and development of social behaviors. Colleges and universities also use this information, to select students from among high school graduates and those having equivalent certificates. The score of the *College Scholastic Ability Test*, administered at the national level, has been the most important measure in this record used in the selection process. Recently the relative importance of the *School Activities Record* within total entrance examination scores has been gradually increased.

Recent reform proposals note that Korean schools have focused on intellectual development in such a way that students failed to develop the attitudes and abilities needed to become responsible citizens. In fourth through ninth grades, the focus will be on democratic citizenship, including rules, processes, and reasonable decision-making. At the high school level, attention will be on global citizenship, including understanding other cultures and peace education. These goals will require new methods of evaluation.

Most assessments have direct impact on classroom learning and teaching. Fundamental instructional changes should occur in response to the goals of enhancing both creativity and compassion in students. Because of this educational reform, assessment methods used in secondary schools will need to be changed. It is clear that a major instrument for change will be assessment. Its role is likely to become one of helping individuals take advantage of the new opportunities for learning created by information technology. Moreover, new learning goals will require the development of new assessment techniques capable of promoting and recording achievement on a very different range of competencies and activities. It is widely recognized that the nature of assessment testing is one of the most powerful determinants of classroom climate and of the content of the curriculum. Assessment can support moves towards the enhancement of active learning in science classrooms.

3.2. *Science curriculum assessment*

Korea has had a national curriculum since 1948. The Ministry of Education in Korea has undertaken revisions approximately every five years in response to the changing needs of society. The development of the seventh national curriculum was launched in 2001 at the secondary school level. Over the past ten years, one of the most distinguishable changes in curriculum reform in Korea is the change of aims including components of science literacy in the science curriculum. It includes a global definition of science literacy. Despite these efforts for the revision of the national curriculum, there has been little change in practice at the school level. To achieve this goal of science curriculum reform, we must attend to many aspects of instruction, such as curriculum content, teaching methods, learning, and assessment. However, above all, we can consider the importance of assessment methods as an integral part of educational reform. Significant curriculum change is not attainable without integral development of new and appropriate models of assessment.

4. ASSESSMENT DATA FOR THE TEACHER

4.1. *Teacher and curriculum*

Teacher conducted assessments should be developed, implemented and interpreted for a variety of reasons. They must be viewed as having several objectives, not simply as a source of grades for students and as a means for establishing their relative ranking in class or school. Assessment is a feedback system to enable teachers to evaluate the effectiveness of the curriculum and their teaching procedures with a view toward changing what is necessary to improve student performance.

- Needed changes in curriculum - If student performance is not as desired or anticipated, several questions must be asked by the teacher. First and most important, were the tests used consistent with the intended learning outcomes? If so then several other questions need to be answered. Are there needed changes in the curriculum? If using a textbook, is it clearly written; are there adequate illustrations; is the information presented appropriate for the educational level of the students? If homework was required, was it consistent with the intended learning outcomes? How was the content of instruction organized? Did it have the potential for effectively presenting information and procedures in science?

- Needed instructional or learning changes - The teacher then needs to address his or her own performance. Were appropriate and effective teaching strategies used? Was the relative importance of the information or procedures communicated effectively to the students? Did the teacher provide sufficient opportunity for students to consult with one another and with the teacher? Did the teacher establish a cooperative and positive learning environment in the classroom? Were field and laboratory experiences used effectively? Were there available community resources, such as museums, universities, agricultural agencies, science research organizations that could have been used to improve students' opportunities to learn the intended outcomes? How effective were community resources if they were used in instruction?

- Administrative support - Was adequate support provide by the school and district administration? Were there opportunities for teacher enhancement activities? Were needs for instructional materials met? Was there encouragement, and financial and personnel support for field trips?

4.2. Data for student diagnosis and advising

The end result of an assessment procedure should not simply be the recording of a grade in a book or other record. It should be used as information for diagnosing each student's progress toward a set of learning outcomes.

- What are the student's interests and abilities? - An effective classroom teacher should be able to discern each student's interests and how they relate to the instructional objectives. Does the student already have certain interests and experiences that, if referred to by the teacher, might help in achieving the intended outcomes? Does a student's interest in science change during instruction? How does such change relate to performance on content or process tests? What background does the student have that might reinforce or detract from performance on a test or in laboratory and fieldwork? Does the student have any learning abilities or disabilities that relate to work in science? Does the student have preferred ways of learning, such as hands-on activities instead of reading or listening?

- Where does the student need help? By answering such questions, the teacher should be able to discern ways in which the student can be assisted in reaching the intended learning outcomes. Are there alternative activities that would be more effective for this particular student than those used in the curriculum? Would the student improve by learning with a group rather then individually, or visa versa?
- In what areas of science is the student successful? A student may not be particularly interested or have the proper background for certain of the intended learning outcomes. Are there alternative areas of inquiry or learning that would be more effective for the student? Areas that would be consistent with the intended learning outcomes.

Here again, using assessment procedures to diagnose student learning problems or advantages is time consuming. The proper use requires the availability of time for both teacher and student, and adequate resources.

5. HIGH STAKES TESTING

Thus far, we have discussed the nature of what has been called formative evaluation at the student and teacher level. This is simply the use of assessment procedures to assist the teacher in modifying curriculum and instruction to suit the learning needs of the student and to diagnose student learning problems and abilities. Perhaps the more familiar aspects of assessment are what have been called 'high stakes testing', especially in the United States where there is considerable attention being paid to standards, and Korea where students are selected for admission to higher levels of education based in large part on their pre-college academic achievement.

5.1. Grading students

Beginning in elementary school, students are often given grades for their performance in science. These grades are based on teacher evaluation of the relative success of the student in knowledge and process acquisition of the objectives of the science curriculum. Teachers assess students in a variety of ways. Too often, however, the major instrument used is an objective test, normally constructed by the teacher from a bank of test items acquired over years of teaching. Such examinations have come under criticism from teachers and other educators as being too limited in their ability to measure some of the more important outcomes of science, such as a student's ability to use science methodology and thinking in decision making. However, properly constructed and evaluated, they can be an important instrument used by the teacher for many of the needs discussed above.

5.2. Evaluation for promotion

Such teacher evaluation can be used to determine whether the student has performed adequately to be promoted; either, (a) advanced to the next grade within a school, or (b) admitted to the next school level (upper secondary, or university)

In many countries, not including the United States, special examinations are given to determine whether a student will go on to a higher school. In Korea for example, teachers are required to give a mid-term and final examination in each of the two yearly terms. These scores then become a part of the student's School Activities Record. This record will be evaluated by the upper secondary school that the student has applied to for entrance. The student must score well on such examinations to be favorably considered for admission to the more academically oriented upper secondary schools. This results in a great deal of pressure on the student to perform well. At the end of the upper secondary school experience examinations prepared under the auspices of the Ministry of Education are given in all of the schools in the country. Good performance on these examinations will help to determine what college or university will admit the student. Finally, in Korea, each university develops examinations that eligible prospective students must take before being considered for final admission to appropriate departments and the university.

5.3. Eligibility for certification (graduation)

In Ohio, the State Legislature has enacted a requirement that students must pass a test, first given in the 10th grade, before they can receive a certificate of high school completion. Fortunately, if they do not pass the test at the 10th grade, they have two years to improve their performance in order to receive high school certification. Similar requirements are being made by other states. However, it has become quite controversial, in part because of the poor quality of the tests and criticisms of some of the content examined.

5.4. Evaluating teacher and school performance

Such test results are also being used to compare schools, school districts and in some cases even teachers as to their 'quality'. In countries where public schools allow movement of students between schools within a district or prefecture, this often results in a hierarchy of schools, with some being favored by parents over others. Often the child who cannot qualify for the top school is economically disadvantaged for the rest of his life, since by not getting into the 'good' high school, neither will he or she be able to get into the 'top' universities where businesses and governments go to get the 'best' talent.

6. ASSESSMENT FACTORS FOR GSL CURRICULA

There are several characteristics of GSL curricula that are either not present in traditional science curricula, or which receive cursory treatment. Although much of what we include in GSL is pretty basic science and can be assessed through measures and techniques used around the world by science teachers, there are several characteristics that are unique. Often it is the context given in the assessment items that may distinguish it from assessment used in traditional types of science courses such as physics, chemistry or Earth science. Often, however, the objectives assessed in GSL curricula will be unique and not normal components of traditional science curricula.

6.1. Internationalization

The emphasis upon internationalization and the understanding and enhancing the appreciation of other cultures through science is one of those unique components of GSL (see figures in Chapter One). Often the teacher will use culture as a context in developing assessment instruments or rubrics. Occasionally, perhaps in consultation with a social studies teacher, the science teacher might want to assess directly his/her students' understanding of cultural elements that may have been at the center of an instructional unit, though in science. In addition, elements of the student's own culture can be used as a context for evaluation of basic science knowledge and methodology.

6.2. Environmental awareness

Enhancing students environmental awareness and knowledge of environmental problems and processes are major objectives of GSL curricula and thus need to be assessed directly by the science teacher (see ESU 2 in Table 1, Chapter One). Fundamental to GSL is the establishment of a sophisticated understanding of the interaction of the various Earth systems and how each might be influenced by human action. Also essential is the development of an ability and interest in following environmental issues beyond the end of formal instruction, and into the student's adult life. Thus, a student's ability to continue to access environmental information over the Internet, in libraries, newspapers and other media should be a component of assessment.

6.3. Using systems methodology

Almost unique to GSL curricula is the emphasis upon the understanding and application of systems science methodologies (see ESU 3 in Table 1, Chapter One). This type of thinking may be of most value to the students as they enter adult life and have to make decisions that cannot be based upon controlled experiments and the type of thinking used by the physical scientist. Therefore, assessment needs to be able to determine the success of activities being used by the teacher in attempting to develop this type of thinking and analysis. Essential, also, is the development of favorable attitudes toward system science. In Chapter One and the Preface, we have mentioned the prejudices that exist in the science establishment that can often be carried over into the pre-college science classroom. Systems science is seldom given equal status with physical science. This may be one of the causes today of the failure of many civic leaders to accept the science of global climate change, pollution, and other environmental problems. Thus, it is especially necessary for the GSL teacher to monitor students' attitudes toward as well as their abilities in, the types of science methodologies.

6.4. Application of physical and biological concepts

In order to completely understand processes going on within the various Earth systems, students must be able to understand and apply many basic physical and biological processes that in traditional science courses are developed through lecture and laboratory experiences (see ESU 4 in Table 1, Chapter One). Thus, the GSL teacher will need to use some of the traditional types of laboratory and conceptual assessment techniques. However, it is important that assessment not just end with the specific physical or biological process. The student must be able to apply that process in the context of an Earth system, and relate how that process functions within that system.

6.5. Use of Internet and electronic databases

An important ability for students in GSL courses is to be able to use the Internet for a variety of information seeking purposes (see ESU 3 in Table 1, Chapter One). At one level it is sufficient to obtain information of the type one might find in a library, only much more up to date. As important, however, is the ability to use the Internet as a source of basic science data that the student can then use under the guidance of a teacher to plot the future path of a hurricane or typhoon, investigate the annual bloom of phytoplankton in different locations along the coastlines of the student's country, or other systems phenomena. Thus teachers will need to monitor students' ability to use such Internet sources and their potential for using them in the future, after they leave formal schooling.

6.6. Earth aesthetics

Science should capitalize on a student's aesthetic feelings about Earth system phenomena (see ESU 1 in Table 1, Chapter One). Thus, teachers need to be able to assess students' aesthetic interests in preparing for class presentations and activities. What art, music or other cultural activities will help to engage students in science? The science teacher can enlist the support of the art and music teachers in identifying materials to use when assigning portfolio activities for example. Rather than the teacher assessing directly a student's growth in art appreciation or the enhancement of a student's appreciation of aesthetic qualities in nature, the purview of the art and music teachers, aesthetics might be used as a context when evaluating student knowledge or process development.

6.7. Understanding very large numbers

The ability to understand great ages and immense distances is not taught to any real extent in traditional science courses. It is a primary objective however in GSL (see ESU's 5 and 6 in Table 1, Chapter One). It is a skill that is essential for understanding and appreciating such major cultural contributions of science as the theory of evolution, plate tectonics, and the immense size of the universe. Therefore the teacher will need to be creative, not only in developing activities to assist students in understanding large numbers, but in assessing their ability to knowledgeably handle the implications of ages of the magnitude of millions and billions of years but also distances of such magnitudes.

7. CONCLUSION

Theories and methods of assessment have been the focus of significant attention for some years now, not only in Korea, but also in many other countries. Curriculum developers have realized that real change will not take place in schools if traditional paper-pencil tests remain unchanged. Assessment can influence and even control teaching and learning. Despite regular educational reform, there has been growing dissatisfaction in Korean education. This criticism of Korean education is not only of teaching and learning, but also of the procedures and technology of assessment.

Korea has a high stakes examination driven education system where formal assessment dominates to the detriment of effective and creative teaching and learning, since most teachers teach to the test. There is a narrowing of the curriculum and pedagogical practice is constrained and sometimes based on outmoded learning theory. The emphasis is on testing for selection and the negative backwash effect on teaching and learning has been profound. This current assessment system raised the problem of 'appropriateness' of assessment, examples of which include a mismatch between the content that is taught and the content that is assessed. Current assessment techniques are seldom appropriate for science learning objectives. We suspect that this is also true in countries that have similar high stakes testing programs, such as Japan, Taiwan, and the People's Republic of China. It may also become a problem in the USA if demands being made for more high stakes testing are met within the various states.

Multiple-choice testing has always predominated in science assessment in Korea. The assessment framework focused on knowledge, understanding and application in science. Paper-pencil testing, using multiple-choice items were the norm, and direct recall items predominated. Recently, there has been growing criticism of the validity of the current methods of educational assessment used in science education. One of the main criticisms is that multiple-choice tests may be a good way for judging students' acquisition of scientific knowledge and information, but they seldom assess, student creativity, problem-solving skills, and higher-order thinking skills. Accordingly, the Ministry of Education introduced performance assessment at all school levels in 1997.

Traditional testing may be the best way to measure some aspects of achievement such as factual recall. In the future, however, the role of assessment is likely to become increasingly one of helping individuals to take advantage of the new opportunities for learning created by information technology. Performance assessments are assumed to tap higher-order thinking processes and be more directly related to what students do in their classroom and what scientists actually do. They are also assumed to be useful policy instruments for science curriculum reform. The theme of performance assessment advocates in Korea has been that these assessments measure important aspects of higher-order thinking. If performance assessments are well designed and include a variety of the objectives and purposes discussed in the chapter, they can provide new conceptions of science teaching for classroom teachers and new information to monitor their teaching. The new educational objectives of understanding cultural and individual diversity, global understanding, creative thinking and problem solving, and environmental understanding and awareness in science implicit in GSL could not be validly tested by paper-pencil tests. Hence, the technical need to develop and use new methods of assessment to be conducted over time within the schools. The adoption of these methods should also alleviate some of the problems with attempts to implement innovative curricula such as those based on Global Science Literacy. At the same time, it is well recognized that traditional testing formats will inhibit such curriculum change, and thus there are also good curriculum development arguments for implementing assessment change concurrent with curriculum change.

SECTION THREE: ISSUES IN STRUCTURING CURRICULUM

In this section several issues in developing and implementing ESE and GSL curricula are discussed. Each chapter is based either on research, or on extensive practice. We trust that the discussion of these beginning efforts and what they have revealed about the feasibility and practicality of developing and implementing new curriculum elements in existing school programs will assist others as similar efforts unfold in various countries.

Nir Orion and his colleagues in Israel have been engaged on a continuing cycle of development, research and implementation of Earth Systems curricula for Israeli middle school science programs. In Chapter Eleven he reviews three lines of research that use the concepts of Earth cycles for structuring curricula. The insights resulting from this research will prove valuable to curriculum developers as they tackle the very difficult tasks of organizing curricula by the concept of the Earth systems instead of the traditional disciplinary organization. Barbara Klemm of the University of Hawaii-Manoa, has had extensive experience writing science curriculum and adaping such for students with special needs. In Chapter Twelve she relies on this experience and her study of the research to describe the potential of Global Science Literacy for serving the needs of these very special children, especially in a camp situation. Roger Trend at Exeter University in Great Britain, has focused his research efforts on the development of the understanding of great lengths of time, popularly called "deep time". This relates closely to Earth Systems Understanding Five (Table 1, Chapter One). In Chapter Thirteen he reviews this research and provides some suggestions to curriculum developers on how to include instruction on this particular understanding, one we believe is essential for citizens' understanding of many of the environmental issues facing the world today.

The last two chapters concern attempts at implementaing aspects of GSL in Japanese schools. In Chapter Fourteen, Masakazu Goto describes his experiences in integrating lower secondary school science curricula in his school, and providing field and aesthetic experiences for his students. He has developed a series of unique approaches that have proven to be effective. In Chapter Fifteen we relate the results of a research study, supported by a Fulbright Research Grant, that sought to determine the feasibility of implementing GSL curricula in Japanese upper secondary schools. The study involved interviewing science teachers in schools in two different areas of Japan. The results are a bit discouraging, not only for implementing GSL curricula, but any type of integrated science curricula. However, the study points up the need for changes at all schooling levels from university at one end of the continuum to elementary at the other, and in areas of public policy, especially the implications for the negative effects of high stakes testing that we found.

Frequently throughout this book, the importance of teacher background and teacher education for implementing new curricula and teaching methods has arisen. There simply has not been sufficient space to examine those challenges in any detail in this volume. A second volume has been developed in which inservice and preservice teacher preparation is a major theme.

V.J. Mayer (ed.), Global Science Literacy, 157.

CHAPTER 11: AN EARTH SYSTEMS CURRICULUM DEVELOPMENT MODEL

Nir Orion, The Weizmann Institute of Science, ISRAEL

1. INTRODUCTION

A reform in science education has grown rapidly during the 1990s in several countries around the world. However, the infusion of the new focus of science education and the new learning and teaching strategies within the educational system still requires much work before it becomes a fully developed way for learning. It is suggested here that the Earth systems approach can serve as an effective holistic framework for science curricula.

The reform of science education in several Western countries is due to three interrelated paradigms, which highly influence the educational systems of those countries:

♦ The "Science for all" paradigm wherein the main goal of science education is to educate our future citizens rather than to prepare them to be future scientists.

♦ The constructivistic paradigm which places the student at the center of the educational process.

♦ The "green" paradigm which emphasizes the awareness of the environment in our daily lives.

These three paradigms are embodied in Global Science Literacy

1.1. The "green" paradigm

As we approached the year 2000, the environmental perspective gained great prominence in Western society. This process, which began two decades ago, has been accelerating in view of the understanding that people's present behavior might actually destroy all of the Earth's ecological systems. Bybee (1993) uses the expression "the orange light has turned red" to demonstrate how serious the problem has become. Approaching the 21st century, people began to internalize their understanding of the significance of their contact with nature. In turn, organizations that call themselves "green" began taking action to preserve the quality of the environment through the development of environment-friendly products, closing factories that cause pollution, and conducting public campaigns against the use of atomic energy.

The call by such groups to preserve the environment and to limit the extent of human damage has been largely addressed to the scientific community, which has been called upon to find solutions to environmental problems. This call to action, as well as the generous allocation of resources, which accompanied it, did not leave the

V.J. Mayer (ed.), Global Science Literacy, 159–168.

scientific community indifferent. Indeed, increasingly scientific research is being carried out in an attempt to understand natural systems and their relationships with human activity. Lovelock (1991) notes that the Earth is composed of several dynamic inter-related systems. He argues that only by developing a multi-dimensional perspective can one understand the global picture. In this light, he proposes that environmental research should be carried out with a multi-disciplinary holistic approach, as opposed to the specific approach, where each scientist specializes in a narrow field that does not relate to the entire picture.

The changes, which have taken place in public opinion towards the environment, may be called "the green revolution". In turn, this revolution has rapidly penetrated the educational systems in the Western world such that the environmental aspect is achieving a central position in the overall system. In GSL it is emphasized through ESE Understandings 1 and 2 (Table 1, Chapter One).

Orion (1996, 1997) proposes that in order to develop environmental insight, the scientific principles interrelated to the environment must be firmly established educationally. Students, who understand their local environment and the processes taking place in it, might know better how to preserve it. In addition, they might have better tools to evaluate the changes taking place in their environment. Similar ideas were suggested by Mayer and Armstrong, (1990), Brody, (1994) and Mayer, (1995).

1.2. The constructivist paradigm

The teaching of science has largely based itself on the paradigm of "instruction", which positioned the teachers and curriculum developers at the center of the educational process. This was in accordance with the perception that students were passive receivers of information that was transmitted directly from the teacher to the student (Bodner, 1986; Bloom, 1956).

Many studies published since the end of the 1970s and onward show the limitations of "instructionism" and proposed an alternate approach known as "constructivism" (Driver, Guesne and Tiberghien, 1985; Osborne and Wittrock, 1985; Bezzi, 1995). The constructivisic approach places the student at the center of educational activity and is therefore also known as the "student-oriented" approach. This approach states that the construction of knowledge is a subjective process and is built by the learner on the basis of his previous experiences. The student perceives the world through the "lenses" of his own personal experiences, which serve as the basis for integrating new information. Thus, learning is not a simple process of adding on information but rather the restructuring of concepts through interaction between the present perception of the student and new information learned. The constructivisic approach relates to learning as a conceptual change, wherein the naive framework which was created in early childhood undergoes modification through learning (Glaserfield, 1987, Gilbert et al, 1982; Driver et al, 1994). This approach argues that since learning is a subjective process, in order to achieve meaningful learning the student must be stimulated and brought to independent understanding. This conclusion has significant implications for the learning and teaching strategies which should be implemented. Fortner in Chapter Five, discussed the "jigsaw" version of cooperative learning as a central learning strategy in GSL.

Active learning is best achieved when students find the content relevant to their world, and when they are left room to feel ownership of the learning.

In this context, students are exposed to environmental issues in their daily lives, since such subjects appear frequently in the mass media. There is no doubt that many of them find subjects like global atmospheric change, air pollution, and water pollution as being relevant to their daily lives. Moreover, destructive environmental phenomena such as floods, hurricanes, earthquakes, volcanic eruptions, land slides, and avalanches attract the interest of many students.

1.3. The "science for all" paradigm

In 1990, the American Association for the Advancement of Science published the policy paper *Science for all Americans* (AAAS, 1990). This document was a part of the AAAS Project 2061, which calls for major reforms in relation to the goals and strategies required for teaching and learning science in schools. The new "Science for All" paradigm perceives the main goal of science education as preparation for the nation's new citizens. In essence, *Science for all Americans* defined minimal levels of scientific literacy for all sciences by outlining objectives for all K-12 students. The *Benchmarks for Science Literacy* (AAAS, 1993) which followed the *Science for all Americans* document, advocated a balance between scientific knowledge, the process of science and the development of personal-social goals (Bybee and Deboer, 1994). In Section One of this book, it is argued that most current efforts, even those initiated by the AAAS come up short in providing the groundwork and guidance to meet this pardigm. We argue that GSL with its inclusiveness of systems science methodlogy and the aesthetics of the Earth system, its emphasis upon the student's habitat as the source of concepts studied, and its social and intra-cultural objectives of global education, provide a comprehensive outline for meeting the requirements of this paradigm. In addition, this new paradigm, which has rapidly influenced many other nations all over the world, gives the Earth systems an equal status among the topics of the science curriculum, in fact it embraces most of those topics and can thus provide a conceptual structure for organizing an "integrated" science curriculum.

2. A HOLISTIC FRAMEWORK FOR THE SCIENCE CURRICULA - EARTH SYSTEMS

According to the above three paradigms, it is suggested that any successfully potential model for a meaningful reform in science education should include the following elements:

- A focus on teaching science in a relevant (daily-life) context.
- The development of environmental insight should be one of the main objectives of science education.
- The design of the curriculum package, and its teaching and learning strategies should be based on a constructivistic cognitive approach.

The starting point of this model is the natural world, which is learned by studying the four Earth systems: geosphere, hydrosphere, atmosphere and biosphere. The study of each subsystem is organized around geochemical and biogeochemical cycles including the rock cycle; the water cycle; the food chain; the carbon cycle; and energy cycles (which are included in all of these cycles). The study of those cycles also emphasizes the interrelations between the different subsystems via transitions of matter and energy from one subsystem to another (based on laws of conservation). Such natural cycles should be discussed within the context of their influence on people's daily lives, rather than being isolated to their specific scientific domains. Humans are introduced in this model as a unique, but integral part of the biosphere. There are many differences, of course, between humans and other organisms; this model emphasizes two of these: 1) The human's ability to produce tools, which is termed technology, and; 2) Their natural curiosity and ability to investigate their environment, which is called science. There is a close relationship between these two characteristics – science and technology – as the progress of one contributes to the development of the other. This model also connects the natural world and technology together, since all raw materials, originate from the Earth systems (mainly from the geosphere and biosphere). The connection between manufactured materials and natural systems of the Earth closes this cyclic model. This connection symbolizes environmental quality, which is profoundly effected by technology. This aspect of the model emphasizes the understanding that society is a part of the Earth's natural system and thus any manipulation in one part of this complex system might adversely effect humans. It is important to note that as opposed to the current models for teaching science, this model does not utilize physics or chemistry as the basis of the curricular sequence. Instead, this model suggests that science studies should start from the concrete world and utilize the concepts and methodology of physics and chemistry as tools for understanding science at a deeper and more abstract level.

The development of such new environmental-based science curricula include the definition of educational goals and objectives. The main educational goal is the development of environmental insight. This can be achieved by learning the following principles:

- We live in a cycling world that is built upon a series of sub-systems (geosphere, hydrosphere, biosphere, and atmosphere) which interact through an exchange of energy and materials;
- Understanding that people are a part of nature, and thus they must act in harmony with its "laws" of cycling. In order to develop environmental literacy, we develop most of our Earth science programs within a systems framework; e.g., *The Rock Cycle; The Water Cycle; The Carbon Cycle.*

Thus we are able to break down the broad subject of environmental insight into a set of more focused research problems whose purposes are:

- To determine the levels of knowledge and understanding of junior high and high school students on the subject of environment.
- To test student understanding of the concepts of cycles and systems, as well to identify misconceptions concerning the Earth's geobiochemical cycles. Moreover we are exploring the deep time concept in relation to the development of environmental insight.

2.1. Examples of curriculum materials

For the last several years, we have been intensively involved in the development of a new Earth science curriculum for the junior high school level. This curriculum is based on the Earth systems approach and includes a unit concerning the hydrosphere for 8th grade students. One of our main goals in teaching in the Earth systems context is the development of environmental insight among the students. This insight is based on the understanding of the cyclic mechanisms of our planet. Our objective is that students should be able to translate environmental problems, such as water pollution into a more coherent understanding of the environment. With such understanding, students might hopefully see the environment as a series of interacting subsystems, that each one is influencing the other.

In order to fulfill the above general goal in the hydrosphere context, we choose the water cycle as the leading concept of the curriculum unit. The curriculum materials present the water cycle as a part of a wider set of recycling systems, which include the geosphere, the biosphere and the atmosphere. The environmental problems are presented in the context of the relationship between the hydrosphere and the other components of the Earth system. Moreover, the relationship between the hydrosphere and the other Earth systems is a result of the transformation of matter (especially water) between them. The development of the new curriculum unit was preceded by a predevelopment study and the implementation of the first development phase was followed by a formative evaluation.

2.1.1. The predevelopment study
The predevelopment study included the following two objectives: a) to identify the previous knowledge and understanding that students hold while entering the junior high school in relation to the various aspects of the water cycle, and b) to explore their perceptions of the cyclic and the systemic nature of the water cycle. In order to collect the needed data a series of research tools were specifically developed for this study. These tools included interviews and open and closed questionnaires. The following is a brief description of the various research tools:
- A questionnaire called *Assessment Students' Knowledge* (ASK): This questionnaire includes two parts: Part A includes a Likert type questionnaire, where students were asked to mark their level of agreement with a list of statements concerning the water cycle. The following are two examples: *1). The composition of a cloud that has formed above the Galilee Lake is different than a cloud that has formed above the Dead Sea. 2).Underground water is actually underground lakes that are located inside the rock.* In Part B, the students were

asked to draw the water cycle on a blank paper. For this task they were provided with a list of the main stages and processes that are included in the water cycle and they were instructed to try to include as much stages and processes as they could from this list.

- A *Cyclic Thinking Questionnaire* (CTQ): In this Likert type questionnaire students were asked to mark their level of agreement with a list of statements concerning the cyclic nature of the hydrosphere and the conservation of matter within the Earth systems. The following are two examples: 1). *The amount of water in the ocean is growing from day to day because rivers are continually flowing into the ocean.* 2). *The cloud is the starting point of the water cycle and the tap at home is its end point.*

- Interviews: Interviews were conducted with 40 students after they answered all the questionnaires. The interview phase had two main objectives. It served us as a tool for validating the students' answers to the questionnaires and it gave us an insight of the students' perceptions of the water cycle. During the interview, students were asked individually to read their answers, first to say if they still agreed with their answer, and then to elaborate on the answer. After the free explanation, the interviewees were asked more questions in relation to their specific explanations.

The population of the predevelopment study included 1,000 junior high school students (7th-9th grades) from 30 classes of six urban schools. The analysis of the different questionnaires of the predevelopment study indicated that most of the students demonstrated an incomplete picture of the water cycle and possessed many misconceptions about it. Children that drew the water cycle usually represented the upper part of the water cycle (processes: evaporation, condensation and rainfall) and ignored the ground water system. More than 50% of the students could not identify components of the ground water system even when they are familiar with the associated terminology. In their mind, underground water is seen as a static, sub-surface lake and water solution chemistry is fixed throughout the entire water cycle. We suggest that those misconceptions reflect students' lack of environmental insight concerning the Earth system.

A significant correlation was found between cyclic thinking and those drawings of the water cycle that included the groundwater component. The following is an example of a student who drew the underground water system and his concept concerning the cyclic nature of the water cycle:

Student: *"I absolutely disagree. There is no starting point and no end point in the water cycle. It is a continuous process".*

The analysis of the predevelopment study findings suggests that the ability of students to perceive the hydrosphere as a coherent system depends on both scientific knowledge and cognitive abilities. The knowledge component is composed of two elements: 1) Factual-based knowledge that includes acquaintance with the components of the water cycle and awareness of its processes, and; 2) Process-based knowledge, namely a deep understanding of the various processes that transform matter within the water cycle. The cognitive component is also composed of two elements: 1) Cyclic thinking, namely the understanding that the water cycle is a system which has no starting or end points and the same matter, but in different forms, is transformed over and over again within the system, and; 2) Systemic thinking, which is the ability to perceive the water cycle in the context of its interrelationship with the other Earth systems.

2.1.2. The development phase

The findings of the predevelopment study served as a basis for the development of an interdisciplinary program named "The Blue Planet". This program focuses on the water cycle as an example of the relationships seen amongst the various Earth systems. It emphasizes a systematic approach by addressing the following aspects:

1. Presenting a coherent depiction of the various processes (chemical, physical, geological and biological) that affect each stage of the water cycle.
2. Relating the water cycle to the different elements of the Earth system.
3. Presenting the water cycle in a Science Technology and Society (STS) format.
4. Using constructivist methods to alter the students' misconceptions of the water cycle.
5. Using computers to access global data bases so those students will better understand that the water cycle is a worldwide phenomenon.

The program also focuses on the role of humans within the water cycle. This was done by including the following subjects:

- Availability of water resources for human use.
- Understanding various components of the water cycle.
- Surface water and ground water resources.
- Human involvement in preserving water quality.
- Understanding Israel's water needs.
- Sustainable development and water resource management.
- Water as an ecosystem.

2.1.3. The evaluation of the first implementation phase

This study examines the effect of studying the water cycle and it's connection with man, on the development of environmental insight among junior high school students. More specifically, we focused on the following aspects:

1. Exploring students' conceptions and attitudes concerning peoples' relationships with the Earth system.
2. Identifying the types of alternative frameworks students possess concerning the various components of the water cycle.
3. Identifying changes in knowledge and cognitive skills developed by students who were exposed to the "The Blue Planet" program.

The research population included 700 junior high school students who studied "The Blue Planet" program. We used the research tools of the predevelopment study modified following the first trail. In addition, we used the following two tools:

- Concept maps. The students were asked to create concept maps in the beginning and the end of the learning process. Comparison of the number and type of items between the concept maps served as a measure of changes in students' knowledge and understanding of processes. The number of connections within the concept map served as an indication of students' understanding of the relationship between the components of the water cycle.
- Observations. In order to track the learning event itself regular observations were conducted in the classes. The observer used a structured observation report that directed her to document the type of activities of both students and the teacher.

2.1.4. Findings
The following are the main findings of the evaluation study of the first implementation phase:

1. Our observations indicated that for the most part, the teachers concentrated on scientific principles and only little on the cognitive aspects of the connections between the water cycle and the other Earth systems, as well as environmental case studies. In addition most of them tended to ignore the constructivistic activities that were specifically developed following the findings of the predevelopment study in order to correct students' misconceptions and to develop a broader and coherent perception of the water cycle within the Earth systems context.
2. A significant improvement was found in student's level of knowledge namely acquaintance with the components of the water cycle.
3. A significant improvement was found in relation to students' understanding of the evaporation process. However, in relation to all the other processes only a minor improvement was found.
4. The analysis of the cyclic and systemic thinking questionnaires (CTQ, STQ) showed some improvements in students' understanding of the different types of interrelationship among the Earth systems. However, even after learning the program, students still had a poor understanding of the systemic nature of the water cycle. Most of the students showed a fragmented perception of the water cycle and make no connections between the atmospheric water cycle and the geospheric underground water cycle.

2.1.5. Conclusion and implication
These findings indicate that the improvement in knowledge is not enough for the development of environmental insight. For this purpose students should develop their cognitive abilities of cyclic and systemic thinking, through learning activities that are directly developed for this purpose. In the current study, we found that although such activities existed teachers tended to ignore them. Thus, greater effort should be invested in teachers preparation and in convincing them that gaining in knowledge does not contribute to a progress in those cognitive aspects, which are critical for the development of an environmental insight.

2.2. The rock cycle and the development of basic scientific skills and cognitive abilities

The rock cycle is a topic included in both the high school curriculum and the new curriculum - *Science and Technology* - for junior high schools. This concept is central because of its great potential for the development of basic scientific skills, such as the ability to identify and to distinguish between observation, hypothesis and conclusion. In addition, it is suggested that it will develop cognitive skills such as cyclic and systemic thinking, which may be crucial components in the development of environmental insight.

In order to be able to read the processes written on the materials of the Earth - rocks, minerals and soils - students are required to understand the different processes, which take place within the Earth. They should also be acquainted with the starting and end products of such processes, and understand that each end product of one process can be a starting product for another process. Such cyclic and systematic thinking abilities require constant organization of knowledge.

This study includes the use of a wide variety of qualitative and quantitative research tools which were specifically developed for this study. Preliminary results indicate that following the learning of an inquiry based learning unit including field trips and an hypermedia software for organizing knowledge, students significantly improved their basic scientific skills. Moreover, the preliminary data suggests effective learning and teaching strategies for the development of cyclic and systemic ways of thinking.

2.3. Developing of environmental insight through the study of the carbon cycle

The Global Carbon Cycle (GCC) is a powerful model for illustrating interactions among various Earth systems and people. This curriculum was developed and implemented in two versions for junior high students and for high school students. The curriculum was built around three sub-units:
1.   The Core Unit: The purpose of this unit was to provide students with a complete and balanced picture of the GCC. An activities guide and periodical guide (which includes a guide to critically understanding scientific readings) were developed for this unit.
2.   Individual Projects: Projects based on the Core Unit and other resources.
3.   Geotop: A field and laboratory unit based upon the Core Unit.

The research was mainly phenomenological and used a wide variety of qualitative and quantitative research tools, which are valid for a small research group. Most of these research tools were specifically developed for this study. Research results indicate that the program contributed to meaningful learning of the GCC. Moreover, it achieved the cognitive goals concerning Earth systems and the role of humans among natural systems. They support the educational effectiveness of using the constructivistic approach and a variety of teaching and learning strategies.

The results indicate a great variance in student knowledge, understanding and ability to perceive the Earth as a whole system. Based on these findings, three critical factors for a whole system perception were identified: perception of interrelationship between sub-systems; the ability to perceive the cycling nature of materials on Earth (also known as the conservation of mass); and a qualitative perception of size, rate and quantity in relation to the Earth systems. The findings regarding the hierarchical structure of whole system perception indicate that this perception is probably not knowledge related.

REFERENCES

American Association for the Advancement of Science. (1990). *Science for All Americans*. New York: Oxford University Press.

American Association for the Advancement of Science. (1993). *Benchmarks for science literacy*. New York: Oxford University Press.

Bezzi, A. (1995). Personal construct psychology and the teaching of petrology at undergraduate level. *Journal of Research in Science Teaching*, 33, 179-204

Bloom, B. S. (1956). Taxonomy of Educational Objectives, *Handbook I: Cognitive domain*, N.Y.: David McKay.

Bodner, G. M. (1986). Constructivisem: A theory of knowledge. *Journal of Chemical Education*, 63, 873-878.

Brody, M.J. (1994). Student science knowledge related to ecological crises. *International Journal of science teaching*, 16, 421-435.

Bybee, R. W. (1993). *Reforming science Education - Social perspectives and personal reflections*. Teachers College Press, Columbia University.

Bybee, R. W and Deboer, G. E. (1994) Research on goals for the science curriculum. In Gabel, D. (Ed.), *Handbook of research in science teaching and learning*. MacMillian, New York.

Driver, R., Guesne, E. and Tiberghien, A. (1985). *Children's ideas in science*. Open University Press, Milton Keynes.

Driver, R., Asoko, H., Leach, J., Mortimer, E. and Scott, P. (1994). Constructing scientific knowledge in classroom. *Educational Researcher*, 23, 5-12.

Gilbert, J., Osborne, R and Fensham, P. (1982). Children's science and its consequences for teaching. *Science Education*, 66, 623-633.

Glasersfeld, E. von. (1987). Learning as a constructive activity. In: Jannet, C. (Ed.). *Problems of representation in teaching and learning of mathematics.*

Lovelock, J. (1991). *Healing Gaia: Practical medicine for the planet*. Harmony books, New York.

Mayer, V. J. (1995). Using the Earth system for integrating the science curriculum. *Science Education*, 79, 375-391.

Mayer, V. J., and Armstrong, R. E. (1990). What every 17-year old should know about planet Earth: The report of a conference of educators and geoscientists. *Science Education*, 74, 155-165.

Osborne, R. and Wittrock, M. (1985). The generative learning model and its implications for learning science. *Studies in Science Education*, 5, 1-14.

Orion, N. (1996). An holistic approach to introduce geoscience into schools: The Israeli model - from practice to theory. In D. A. Stow and G.J. McCall (Eds.), *Geosciences education and training in schools and universities, for industry and public awareness*. AGID special publication series No 19. Rotterdam: Balkema.

Orion, N. (1998). Earth sciences education + environmental education = Earth systems education. In Fortner, R. W. and V.J. Mayer, (Eds), *Learning about the Earth as a system. Proceedings of the Second International Conference on Geoscience Education (pp. 134-137)*. Columbus, OH: Earth Systems Education, The Ohio State University. ERIC Document ED422163.

CHAPTER 12: ENABLING GLOBAL SCIENCE LITERACY FOR ALL

E. Barbara Klemm
University of Hawaii-Manoa, USA

1. INTRODUCTION

Commitment to the inclusion of learners with disabilities is a necessary, integral part of making Global Science Learning "for all." However, to do so requires two changes: transforming our thinking about individuals with disabilities from focus on their disabilities to focus on their abilities and modifying learning environments so that they are inclusive, enabling, and accessible for all learners.

2. BACKGROUND

Late this afternoon we flew kites. Neither my buddy nor I had ever flown a kite before. Before we tried, we watched what others did, running with their own kites to launch them into the air. Paul used his powerful arms to toss the kite into the air, and then hang onto the kite string; yet, somehow, he also helped me while I started running and pushing his wheelchair across part of the grassy field. Our kite faltered at first, but soon it lifted and Paul fed out the line and hung on as the kite strained upward. I stopped running and we both stared in delight as we watched our kite join the others, way up in the sky. I suspect that the obvious delight on Paul's face was mirrored on my own. We learned how to fly a kite, and I learned from Paul lessons that I am still pondering. (Klemm, personal journal, Sept.1999).

When I wrote that journal entry, I was a volunteer adult leader in one of several three to four-day science camps for adolescents with disabilities that are described later in this chapter. Earlier the same day Paul (a real person, but with a different name) had asked me "Do you mean that I can do something like that too?" Then, we were not talking about flying kites; we were talking about possibilities for Paul to pursue a career as an aquatic biologist, so that he, too, could work in the area of wildlife conservation. That morning Paul and two dozen other adolescents with various moderate to severe disabilities had interacted with a field biologist at a beach park near the campsite. We engaged in hands-on opportunities to set and "read" monitoring devices and to handle live fish, turtles and crustaceans caught for observation. Paul, whose disability rendered him paraplegic, had never thought about such possibilities for himself.

As we envision Global Science Literacy (GSL), what possibilities do we educators hold for our students—for all our students, including students like Paul?

V.J. Mayer (ed.), Global Science Literacy, 169–185.

Do we have a conceptual and pedagogical foundation to support "GSL for all"? Who is included in "for all"? What do we need to do to make GSL fully accessible to all students, including those with disabilities? To provide a basis for discussing "global science literacy for all," I draw upon education experiences from the United States to provide both the language and a sense of evolving understanding of what educators in the U. S. mean when they claim that education is "for all."

2.1. Education Standards

GSL has foundations in both science and environmental education. What do these areas say about "for all"? *Science for All Americans* advocates a common core of learning for all young people "regardless of their social circumstances and career aspirations," particularly pertaining "to those who have in the past largely been bypassed in science and mathematics education: ethnic and language minorities and girls," (Rutherford and Ahlgren, 1990, pp. ix-xi). The subsequent *Benchmarks for Science Literacy* (American Association for the Advancement of Science, 1993) and *National Science Education Standards* (National Research Council, 1996) also speak to the need for "science literacy for all," this time acknowledging "special needs" students. They explain their stance that "science is for all students" as follows:

> This principle is one of equity and excellence. Science in our schools must be for all students: All students, regardless of age, sex, cultural or ethnic background, disabilities, aspirations, or interest and motivation in science, should have the opportunity to attain high levels of scientific inquiry. (NRC, 1996, p. 20)

When the North American Association for Environmental Education formulated its guidelines, it expanded the notion of "for all" as including all social groups, ages (calling for intergenerational learning), and geographic differences (NAAEE, 1998). The NAAEE guidelines include those involved in both formal (school-based) and nonformal environmental education. The notion of "for all" in the National Science Education standards assumes that "all students are included in challenging science learning opportunities," emphatically rejecting any situation where some people in certain populations are discouraged from pursuing science and excluded from opportunities to learn science (NRC, 1996).

2.2. U. S. Federal Laws

In the United States, education is governed at the state level, and most U. S. public schools are funded locally through property tax revenues. However, federal laws enacted to protect rights guaranteed to all U. S. citizens now shape education policy and practice in all states. In particular, civil rights laws pertaining to educational equity have had far-reaching impact on U. S. education (Banks, 200; Lee, 1999). Over the past half-century, historical, political, and social developments nationwide brought about the passage of federal laws aimed at educational equity, including desegregation, affirmative action, gender equality, and bilingual education. The 1954 Supreme Court decision in Brown v. the Board of Education of Topeka set the stage for the Civil Rights movement of the 1960s and 1970s, making educational equality a central component of educational discourse in the U. S. (Banks, 2000). In 1965, educational equality was construed in terms of economic equality, so students in urban and rural schools where poverty prevailed were considered economically disadvantaged. Passage of the Elementary and Secondary Education Act (ESEA) in 1965, specifically under Title I of the act, channeled federal aid to where test scores are lowest and poverty is highest, the latter determined in part by the percentage of students receiving subsidized lunches. ESEA remains in force today, providing sorely needed funds to urban and rural schools in economically depressed areas. ESEA also supports such programs as Title VII Bilingual Education materials and staff development, viewed as a national need to empower schools to meet needs of language-diverse students.

Pipho (2000) believes that "court decisions on topics such as desegregation and finance equity have had a stronger impact on education than research studies." Even though federal funds constitute only about six percent of the overall school budget nationwide in the U. S., threat of loss of the ESEA federal funds drives school compliance with provisions of federal laws addressing minorities and women, and more recently, persons with disabilities.

2.3. Students with Special Needs

The U. S. Congress enacted several laws, mandating that all students, including special education students, have access to the general curriculum. These laws:
1. Guaranteed basic civil rights to people with disabilities (Rehabilitation Act, 1973, Section 504); and guaranteed people with disabilities access to all walks of life (Americans with Disabilities Act of 1990, P. L. 101-336).
2. Provided free, appropriate public education for all handicapped children; that children are taught in a least restrictive environment; and that an individualized education plan is developed for each identified child. (1975, P. L. 94-142).
3. Included infants and toddlers, required individualized family service plans, and suggested individualized transition plans for adolescents with disabilities (1986, PL 99-457)'
4. Changed the language of the law from "handicapped children" to "children with disabilities"; included autism and traumatic brain injury; and clarified that states may be sued in federal courts for violating the laws (1990, P. L. 101-476.
5. Provided access for children with disabilities to the general curriculum and to participation in state and district-wide assessments; involved regular education teachers; and provided for graduation with a regular diploma (1997, P. L. 105-17).

Traditionally, students with special needs have not had the same access to general curriculum as other, non-disabled students. Instead of educating students with disabilities separately in homogeneous, self-contained special education classrooms where they have little interaction with other students, special education laws require that appropriate services be provided for students with disabilities, so that they can learn. They ask educators to modify their teaching to include students with disabilities with their peers in the general curriculum. The general education curriculum includes the traditional subject areas, consisting of the core curriculum of English, social studies, science and mathematics; electives such as the arts, vocational education and physical education; and those integrated subject areas like environmental education. Thus, special education laws affect all educators at all levels (pre-college or college), and also apply to non-formal education settings, such as museums, parks and nature centers.

Special education students are those identified as requiring special adaptations in order to learn, as defined under IDEA 97, the re-authorized Americans With Disabilities Act, and Section 504 of the Rehabilitation Act of 1973. Nationwide, approximately 10-11% of the entire school age student population is considered disabled, although there may be more with undiagnosed learning disabilities. Incidences of disabilities reported by the U. S. Department of Education (1998) indicated that 90% of disabilities were classified in one of four categories: Learning Disabled, 51%; Speech or Language Impairment, 21%; Mental Retardation, 11%; or Serious Emotional Disturbance, 9%. The remaining 10% of low-incidence disabilities varied among eight categories: Multiple Disabilities, 1.9%; Hearing Impairments, 1.3%; Orthopedic Impairments, 1.3%; Autism, 0.7%; Visual Impairments, 0.5%; Traumatic Brain Injury, 0.2%; Deaf-Blindness, 0.03% and Other Health Impairments, 3.1% (USDOE, 1998).

All students with disabilities require services to help them learn, communicate, and behave appropriately in school settings. Relatively few students with disabilities have readily apparent, severe physical impairments. Most students diagnosed with disability appear, at first, as indistinguishable from their peers, but unlike their peers, they need special adaptations in order to learn.

Thus, in the U. S., special education is mandated by law and governed by legislation, court cases and regulations, but none of these address the subject matter content to be covered. Special education rests on law and is not a subject area, so it has no voluntary national content standards, as do the traditional subject areas of mathematics, English, social studies, or science (Klemm and Avery, 1999).

Special education laws address the learning needs of individuals with disabilities. Educational reform under these laws is specific to each individual diagnosed as requiring special adaptations in order to learn (IDEA, 1997). Therefore, special education laws are changing U.S. education at the targeted level of one learner with disabilities at a time. What was once probably unthinkable— that some students receive special accommodations, such as an alternative way to take a test, or more time to complete the test—is now mandated under special education law. If such accommodations seem problematic, consider what a special educator explained to this author, that providing specific accommodations for learners according to diagnosed disabilities is like permitting students with poor eyesight to wear reading glasses when taking a test.

Special education laws are bringing about structural changes in education because the laws state that children with disabilities are to be educated in the "least restrictive environment." Now, an Individual Education Plan (IEP) must be developed to uniquely suit each child with disability enrolled in public schools (1975 P. L. 94-142). One difficulty arises because content and coursework traditionally are thought of in terms of grade-levels, as is evident in the *National*

Science Education Standards (NRC, 1996) that are arranged in three grade-level clusters, Kindergarten-4, 3-8, and 9-12. Grade-level and implied age-level designations for content learning can be problematic within the framework of the special education laws, which focus on the developmental level of each learner with disabilities. For example, an adolescent learner who functions at a "third grade" reading level is unlikely to be interested in the content of reading materials appropriate for third-grade aged children. Each IEP also takes into consideration the learner's age, functional skills, and the prerequisites needed for learning (Council for Exceptional Children, 1999). For severe disabilities, the IEP focuses on that individual learner's needs to develop functional life skills, such as the ability to communicate or use a telephone.

In practice, children with mild to moderate disabilities are being placed with their non-disabled peers into general education classrooms, a practice called inclusion or mainstreaming. Under the special education laws, placement into general education classes is determined by the needs of the child in terms of gaining functional skills, such as speech or communications, and by what various classes offer in terms of helping that child attain personal learning goals. The special education teams who develop the IEPs include school counselors, parents, teachers and others. For adolescents with disabilities, course placement is determined by selecting those courses that best fit the needs of each learner with disability. Special educators favor classes using active, hands-on learning approaches for integrating students with disabilities, particularly those found in science (Scruggs and Mastropieri, 1993; Siegel-Causey, 1998).

Practical and proven instructional strategies for teaching science to students with disabilities in general education settings include (a) instruction that builds on learners' prior knowledge and connects to their everyday lives (Choate, 2000; Kame'enui and Simmons, 1999); (b) curriculum that offers opportunities for different modes of teaching and learning such as multisensory (visual, auditory and tactile) experiences, and lessons that address differences in learning styles and multiple intelligences (Gardner, Korhnaber and Wake, 1996); and (c) "conspicuous strategies" for instruction, such as using visual graphics, demonstrations, videos, role-playing, simulations, and manipulatives (Rakes and Choate, 1997; Vaughn, Bos and Schumm, 1997). Other practical and proven strategies include use of guiding questions, and emphasizing process skills and problem solving approaches.

2.4. Enablement for All

Arguably, U. S. federal laws on disabilities benefit everyone. Today, for example, the public benefits from ramps constructed for wheelchairs because the ramps also accommodate parents pushing children in strollers. Also, restrooms in public locations are now modified to accommodate the elderly disabled, as well as the young; and commercial businesses now provide parking spaces and shopping accommodations for the disabled, for pregnant women, and for parents with young children. The laws have acted to change physical, public spaces so that they are more accommodating and enabling for all.

Children with disabilities and their families are clearly the targeted beneficiaries of the special education laws. Even though only a small percentage (about 1%) of students have physical disabilities, classrooms and laboratories must be physically accessible, so new or reconstructed classrooms provide such accommodations as lowered laboratory workbenches for wheelchair accessibility. As the statistics on the incidence of disability show, only a small percentage of children with disabilities require specialized accommodations such as computer devices that translate text into large letters or into Braille, and special listening or recording devices. The laws also make provisions for special teacher aids if needed, such as requiring schools to hire sign-language interpreters to assist learners with profound hearing impairment.

The special education community has transformed its thinking from a "paradigm of disability" into a celebration of diversity and enablement. The traditional view of disability is negative, looking at the deficits or limitations of handicapped individuals. The current view is a positive, ecological view of disabilities as not inherent in an individual but, rather, viewed as limitations to full participation resulting from inaccessible physical environments, and lack of accommodation, support, or inclusion in social environments (National Institute on Disability and Rehabilitation Research, 1998).

This newer, ecological vision of diversity and enablement advocated by the special education community supports another shift in perspective taking place among the general education community that is sometimes called "focus on the learner." As Williams (1999) explains, focus on learners and learning involves a subtle shift in educational thinking from teaching to learning and from an information-transmission view of traditional learning environments to a "constructivist/cultural" view. Making learning accessible to all learners, with or without disabilities, necessitates the use of "best practices." Pedagogy that embraces diversity and strives to make learning fully accessible to all learners fosters group cooperation and peers supporting one another so that each succeeds and learns. Accessibility and educational enablement mean using research-based best practices, including the following:

1.   Cooperative learning. Peer learners working together to achieve common objectives and to help each other succeed is not only central to cooperative learning, but also essential for classroom and field experiences that include students with disabilities. The special education notion of enabling environments reminds us that learning situations can be structured to support learners' contributing based on their abilities, not disabilities.
2.   Inquiry teaching and learning. For example, a basic inquiry planning model familiar to most science teachers is the Learning Cycle and its 5E modification: student engagement, exploration, explanation, elaboration, and assessment.
3.   Questioning strategies that support critical thinking and constructivist approaches to learning. An important skill for teachers is the ability to scaffold responses, particularly in tailoring questions so that they are not so open-ended as to frustrate a learner, nor so narrow and specific that they undermine inquiry, but just right in terms of supporting individual learners in actively thinking and constructing understanding.

4.   Authentic assessment and alternative assessment formats. Science education standards today call for assessment to inform teaching and learning, and so achievement and opportunity to learn science must be addressed. Assessment is authentic when assessment tasks "are similar in form to tasks in which they [students] will engage in their lives outside the classroom or are similar to the activities of scientists" (NRC, 1996, p. 83). Classroom assessments today take many forms, including performance-based observations as well as effective traditional tests and examinations.

5.   Respect for learning diversity. Cooperative learning, inquiry, scaffolding and multiple intelligences contribute to an educational climate that respects learning diversity. Especially in hands-on inquiry teaching, lessons today provide opportunities for addressing learning differences, including preferred learning modalities (visual, auditory, tactile) and multiple intelligences.

3. EXPERIENCE USING ESE UNDERSTANDINGS FOR ALL STUDENTS

The conceptual understandings for Earth Systems Education (ESE) can—and should—be implemented as a framework for all learners, including students with special needs. To support this contention and illustrate how each ESE understanding can support learners with disabilities, I draw examples from my experiences with the Ocean of Potentiality Marine Science Camps (OP) for adolescents with disabilities, funded by the National Science Foundation and other sources. The purpose and design of the OP camps is discussed more fully elsewhere (Klemm, 1999; Klemm, Radtke, and Skouge, 2000; Klemm, Skouge, Radtke and Laszlo, 2000).

Paul, mentioned at the beginning of this chapter, was one of about two-dozen adolescent campers in one of the OP camps. The camps bring together adolescents from all islands in Hawaii, with a spectrum of moderate to severe disabilities: cognitive, emotional, hearing, sight, speech, and orthopedic impairments. To the best of my knowledge, this diversity makes OP camps unique. Camps elsewhere either include few students who are disabled, or are designed for a specific type of disability, as for example, a camp for hearing impaired and deaf youth. The OP camps are also highly diverse in other ways: they include male and female youth and adults from many ethnic backgrounds (reflecting the rich cultural pluralism of Hawaii); youth are invited from urban and rural areas of the state; and intergenerational activities engage youth with adults, who range from young adults to senior citizens.

Adults, including disabled and nondisabled scientists, parents, special educators, science teacher educators, science teachers and pre-service teacher candidates, assist as needed in making OP camp activities accessible to all participants. To provide structure, campers work together in cooperative teams, each composed of youth with a variety of disabilities plus two or more adult camp leaders. Each person is expected to participate, and to help others participate as well. Diversity is celebrated—each person is different and everyone has abilities. Importantly, the camps are not part of the formal school program, which by law must create an Individual Education Plan (IEP) for each child. Therefore, the nonformal OP camps are free to try new approaches for making learning accessible to all. Instead of pouring energy into creating IEPs for each learner, camp leaders focus on devising ways to enable all youth with disabilities to engage in camp activities. As a science educator, I volunteer, in part, to have the opportunity to work with others to create enabling learning environments, to participate with the youth and others during the camps, and to learn from the campers who are disabled how make science more accessible for all students.

3.1. Understanding 1. Earth is Unique, a Planet of Rare Beauty and Great Value

This ESE understanding invites students to learn about the Earth through aesthetic experiences. For classroom settings, Mayer suggests using art, music and literature to engage students aesthetically in thinking about planet Earth (Mayer, 1989). Rather than using aesthetic examples created by others, OP science camps provide opportunities for the youth campers to create their own aesthetic expressions, using a mix of traditional expressive technologies plus digital multi-media technologies. Each OP camp provides opportunities for participants to explore several types of natural environments. Depending on the camp location, we have explored tide pools and sandy beaches, tropical rainforests, streams, and dry, steep-sided valleys. Wherever we go, we encourage campers to use all their senses (sight, sound, touch, and feel), and to engage both their right brain (feelings) and left-brain (logic) to "observe" and "experience" the environment.

For years, I have espoused direct, experiential learning, and by that I have meant, literally, hands-on, multi-sensory learning experiences that engage students in actively thinking about and investigating the world, and in constructing their own understandings based on these experiences. From my OP camp experiences, I have learned that digital technologies, computers and other forms of "high technology" also provide powerful sensory augmentation, as well as generational-appropriate alternatives to traditional written and artistic media for expressing ideas about the natural world.

During field activities, OP campers have access to video camcorders, digital cameras, Polaroid cameras, and small audio recorders; at the campsite, access to computers, software, printers and video playback equipment. We ask each camper to keep a daily journal and produce a camp memory book. Campers have time in the evenings to reflect on their experiences, including opportunities to view and talk about their video and audio recordings of their experiences, to write and illustrate their journals, and to compile their camp memory books.

Rather than requiring specific observations or specifying the format for recording and reporting experiences (i.e., provide guidelines for a "field trip report," each team is free to select its own form(s) of expression, and to share in use of the available multi-media equipment. The campers quickly learn how to use the laptop computers and software to sort, edit and narrate their work. When needed, instruction in the use of equipment is individualized, often with youth helping adults, as well as one another.

We do require each camper to create a personal "camp memory book" to take home, in which they record daily journal entries and illustrate their experiences. During our first camp, this consisted of a three-ring binder, paper, pencils and pens, all traditional "low-technology" forms of note taking. To invite creativity, and make the process more enjoyable and the product visually more pleasing, we now also provide colored soft-cover binders; brightly colored pens and pencils; brightly-colored paper, some with designs; regular scissors and specialty scissors with sculpted tines (e. g., to cut out a scalloped or zigzag border); a variety of rubber stamps with environmental designs, plus different colored stamp pads; an assortment of stickers; glue; and even glitter. Students decorate the cover of their camp memory book, and make daily entries using whatever materials they choose. They can add digital images and computer-processed texts and graphics to their paperbound, hard copy pages. We began experimenting with modifying an electronic portfolio approach for campers to produce an electronic camp memory book, and, after the camp, posting camp experiences on an OP networking web site.

Students with learning difficulties require more guided structure. We provided prompts for the daily journal entries, encouraging reflections on feelings as well as cognitive learning. Some examples are: "Today I wondered..."; "The most beautiful thing today...."; and "Something I learned that I never knew before...." Not incidentally, OP camp planners do prepare resource materials for campers to use and to place into their camp books, including daily schedules, written and illustrated activity guides and additional materials, such as a tide chart and maps of areas explored during camp activities. These materials are intended as supplementary resources, not as a text-based means for engaging campers in activities.

All campers are engaged, and each not only participates, but also contributes in some fashion. Although the products and performances of the campers could surely be considered forms of authentic assessment, instead of evaluating the outcomes, we reward the effort, mostly using praise and group recognition. (Evaluators formally assess the camps themselves, providing formative feedback for planning the next camp.) The resulting freedom opens opportunities for campers with disabilities. One student with hearing impairment used earphones and a video-camcorder, electing to be the camp "roving reporter" supported by others in his team who planned to "produce" a "documentary television show." Another student worked by himself, occasionally seeking advice from an adult leader on elements of graphic design, and, in the end, produced thoughtful, aesthetic digital portraits of the camps social and natural environment. Some teams managed to develop the beginnings of a credible "story-line" to present their experiences; others prepared collections of images, more like a collage of the OP camp activities.

The tried-and-true concrete, hands-on approaches still work, too, as I saw with a camper with severe cognitive impairment, who was enthralled in making paper pin-wheels ("wind-mills") and releasing soap bubbles to detect air movement, direction and approximate speed. He was especially delighted to have a digital image of himself in action to add to his camp memory book. Thus, learning in an OP camp is less structured than in a classroom. However, the camps provide a supportive and enabling environment so that all campers engage in activities and share their experiences with others, using multi-media forms of expression, which they choose for themselves. Our camps demonstrate the power of the first ESE understanding as an entry point for engaging all learners in studies of the world around them.

3.2. Understanding 2. Human Activities, Collective and Individual, Conscious and Inadvertent, are Seriously Impacting the Earth

During these camps, we do not formally require campers to make observations about the impact of human activities. However, we do require the campers to participate in one or more service projects to "give back" to the community and the environment. Service and 'giving back' are important for students with disabilities, especially those with severe disabilities, who in everyday life are the recipients of care giving. Unlike their nondisabled peers, youth with disabilities have limited opportunities to engage in outdoor explorations, and, arguably, very few opportunities to actively contribute in volunteer service. During one of our camps, all students, even those in wheel chairs, participated in a beach cleanup activity. Working together, we picked up beach debris, sorted and classified it, and recorded our findings on forms provided by a government agency that sponsors beach cleanups. The activity provided a context to talk about actions that led to the beach's despoilment, and to ponder our own behaviors and responsibilities.

Although the OP camps have not formally designated "place" as a theme for each camp, campers engage in exploring the natural and human history associated with the locations of each camp. For example, during a camp held in a dry leeward valley, we learned about traditional Hawaiian and modern-day views on water rights. Our purpose is to introduce campers to several such ideas during the camp, rather than focusing on a single issue. THUS, OP camps facilitate awareness building, but do not go much beyond initial exploration in the inquiry process.

3.3. Understanding 3. The Development of Scientific Thinking and Technology Increases Our Ability to Understand and Utilize Earth and Space

OP camps seek to engage campers in authentic science experiences, with these experiences conducted out-of-doors whenever possible. We select OP camps at various locations in relatively remote locations, but where facilities are handicapped accessible, such as having bathrooms with at least one toilet and shower that are wheelchair accessible. We then plan the campsite and field activities according to the location of the camp, including how to make these activities accessible to all OP campers. Logistically, OP camps require extraordinary planning, as well as the support of many caring volunteers, especially to provide necessary support for the campers who require physical or additional learning assistance. We made sure that an adult companion accompanied a camper who is blind, and we provided additional visual instructional aids for those with profound hearing disabilities (in addition to the services of an American sign-language interpreter). Some of the campers require round-the-clock assistance in order to eat, drink or use toilet facilities. We also had to provide rest periods for the volunteers themselves. So, all told, we need about one or more adult volunteers for each camper, depending on the nature of the disabilities of the youth at the camps. Each adult is assigned one or two youth camper buddies, two or three adult volunteers share in facilitating youth teams, and camp leadership responsibilities are divided among the camp planners.

OP camps are an example of an *ad hoc* field experience, in that each is held at a different site and planned for a single event. OP camps differ from most other formal or nonformal education settings, where far fewer learners with disabilities are congregated together. Most other school or camp events are held at established sites, and most offer programs that are repeated periodically. OP differs from school even where students often know each other and learning takes place over weeks or months. Now that we have conducted six OP science camps, we now have a cadre of experienced planners and adult volunteers, but as long as each camp is held in a different setting and for students with varying disabilities, each camp will be planned anew.

Because most of the OP campers have had little or no prior experience with field studies, they are novices in field situations. Any novice, whether disabled or not, observes obvious things, but notices few details or changes, has poor discrimination ability, and may not use-or know how to use-all of his or her senses (Southwest Consortium for the Improvement of Mathematics and Science Teaching, 1996). For OP campers, and for any novice in field observation situations, we have to provide enough enabling background information to orient them. With or without disabilities, any novice observer, who has never "really seen" or been in a tide pool, needs to do just that—take some time to be there, to take in sensory impressions, and to begin noticing details. Until novices have had some initial experience, they are not ready to be rushed into observations that are more systematic. They need to see the organisms in a tide pool first, for example, before engaging in something like laying a transect line and conducting quadrant sampling counts of organisms. Learning scientific thinking skills is a developmental process, and so our greatest constraint is time, not the disabilities of the campers.

As we would for all novice learners, we use interactive learning strategies that are well known to outdoor environmental educators. These include prompting the campers to use all their senses, i.e., hearing, touch, smell, and sometimes taste, and encouraging them to use all their learning abilities, i.e., expressing their feelings as well as their cognitive thoughts. We provided additional learning support intended to help campers use their abilities. For campers with visual impairment, we provide tactile examples; and for campers with hearing impairment, in addition to their American sign-language interpreter, we provide visual instructional aids. Situationally, we respond as needed for campers with motor disabilities, as for example, holding a cup-and-string "telephone" to the ear of a camper unable to move his arms, or making sure that those in wheelchairs are positioned where they, too, can view events. Most of the adult volunteers have little background in science or science inquiry teaching, so they are participant learners, as well as facilitators. Youth campers help each other, providing invaluable supportive care and encouragement for one another. As a result, everyone participates.

Not only for campers with cognitive disabilities, but also for all campers, we made an effort to employ instructional "best practices." For example, we use multi-modal instruction, meaning that we give prepared written guiding materials, orally go over the hand outs, post summarized key ideas on charts, and visually point to specific items on the chart. Whenever possible, we demonstrate with specific, concrete examples. Rather than focusing on accommodations for individual disabilities, we focus on creating an inviting and supportive environment for all the campers.

Of all the categories of disabilities, I personally was most concerned about working with campers who are behaviorally disturbed. Based on my own experiences as a teacher in active, hands-on science classrooms, students with persistent behavior problems challenged my teaching skills the most, usually because they are disruptive. However, the OP campers proved themselves enthusiastic and appreciative. They did not disrupt activities. Seemingly, the campers with behavioral disturbances flourished in the less formal camp setting.

Throughout the camps, the special educators, in particular, made clear that the campers had come together like a family, and that they were expected to respect and care for each other, and to conduct themselves in a way that would make everyone proud of the camp. We explain that we are all part of an *ohana,* the Hawaiian term with cultural and behavioral relevancy that encompasses the notion of being embraced by an extended family, with attendant responsibilities. We also follow through with positive reinforcements. For example, each evening, we asked each camper, including the adults, to publicly thank or acknowledge someone for his or her efforts, including helping others, and personally risking to experience new things. We make the positive reinforcements tangible, too. As campers acknowledge each other, the person recognized received a sticker to wear on his or her tee shirt. Of course, the adults joined in too, mostly to make sure that everyone was mentioned and to give our youthful "buddies" special recognition.

If the OP camps are to provide opportunities to develop scientific thinking, then they must make sure that "enablement" does not diminish inquiry. A challenge in working with learners with disabilities is to strike an effective balance between stepping in to provide appropriate support, and stepping back to allow the campers to observe, think and act on their own, or with peers.

Just as scientists use technology to understand the world, so too is the use of technologies a part of the OP camp activities. We use digital technologies to observe and report experiences, as described earlier. We also use authentic techniques, but inexpensive equipment. For example, we pulled wire coat hangers open to form quadrants, which we used for a transect study. Visiting scientists brought authentic equipment too, as for example, the aquatic biologist who brought along his stream monitoring devices. For our most recent camp, we purchased hand-held global positioning devices and have begun to learn how to use them with map activities.

Some of the campers use augmentative and assistive technologies to transcend their disabilities. One camper used a talker, a computer keyboard-like device that produces comprehensible audible speech; and another used a Braille device, that makes printed materials readable for the Blind. Highly trained, personal assistance dogs have also been part of each camp. Some dogs are "seeing-eye" guide dogs for the blind. Others are service dogs, trained to respond to the commands of someone who is paralyzed, such as fetching human help or activating the power button on a motorized wheelchair. We encourage the campers with enabling devices to demonstrate and explain them to all of us. We also ask the campers with trained dogs to teach us how to act when a dog is "on duty" and trained to be alert, and during its rest periods, when the dog is feed and groomed. This setting, where all the youth are disabled, helps the campers overcome shyness and feel at ease with each other. Too often, in other settings, youth with obvious disabilities do not want to draw attention to themselves, or they may become isolated because others are uncomfortable in interacting with them.

3.4. Understanding 4. The Earth system is composed of the interacting subsystems of water, solid Earth, ice, air and life

The camps provide wonderful opportunities for firsthand observations of Earth systems, including volcanic island formation, erosion, the water cycle, and relationships between rainfall and vegetation. Our limitation for involving OP campers in such investigations is not the disabilities of the youth, but the short duration of the camps, coupled with the inexperience of many of the youth in camp or outdoor settings. Thus, we plan strategies to maximize learning opportunities and to promote at least awareness-level understanding of Earth systems. For example, both before the camps and during campsite activities, we assign specific topics to campers, charging them to prepare short presentations to give to their peers. In doing so, we more actively involve campers in their own learning, and foster interaction. We recognize that planning a presentation is a challenge, and giving a presentation in front of peers, a potential social risk. A camper disabled with partial brain damage and a resulting speech impediment, for example, made a poster showing the water cycle, which she explained to her peers. Later, in the field, knowledgeable adults helped everyone to observe specific components of the water cycle, such as clouds forming on the windward side, waterfalls, and valleys formed by stream erosion. As we did this, we made sure that we referred to the poster presentation of this one camper, thereby further acknowledging her effort and contribution.

Although we do not formally sequence learning activities in the camps, experiences are loosely linked. For example, at the camp where we looked at the water cycle, we also flew kites, released soap bubbles and made paper pin-wheel (toy) anemometers. We informally coupled these hands-on activities with field observations of the direction and speed of wind, cloud formations, rainfall patterns, land erosion, and evidence of the water cycle.

3.5. Understanding 5. Planet Earth is more than 4 billion years old and its subsystems are continually evolving.

The main Hawaiian Islands where we live are at most under two million years old, and some of the volcanoes that are building the islands are among the more active in the world. Although we could readily do so, we have not yet developed activities for the OP camps focused on the geological timeline for evolution of the Hawaiian Islands. We have looked at evidence of weathering and erosion, of the rise and fall of sea level (as visible in the various coral benches). As I write this, I can readily envision future OP camps where we enable all of our campers with disabilities to see and experience the remarkable geological features of the islands. We are fortunate that these are right here, and readily observable. I am also aware of simulation activities to help learners understand the enormity of timelines measured in billions of years. Understanding 5 has served, for me, the purpose of reminding me that this, too, is part of Earth systems science.

3.6. Understanding 6. Earth is a small subsystem of a solar system within the vast and ancient universe

To date, the OP camps have not emphasized this understanding as such. However, on cloudless nights, we encourage campers to observe the moon phase and look for the constellations. A number of our campers have had little prior stargazing experience, especially away from urban lights. As an aside, I wonder how many urban students anywhere today lack such experience.

Explaining what we observe is another matter. As a science educator, I am prepared to help novices begin to observe stars as I learned to do so, from a science perspective and from some navigational experience. My understanding of Polynesian way finding and ways of knowing is limited. Although we have invited Hawaiian storytellers to the camps, they have told us scary ghost stories and tales of their childhood days, before jet planes, satellite communications, or even electricity. Such firsthand inter-generational stories, I believe, help all of us comprehend human history, and especially how technologies have transformed us. But, our camps—and the GSL movement as a whole—needs to learn more about non-Western, traditional ways of explaining the world and the ancient universe, too. The holistic stance of Global Literacy can serve to bring together understanding of what is now divided into science, traditional ecological knowledge, history, art, music, literature, religious perspectives, and multi-cultural customs.

3.7. Understanding 7. There are many people with careers that involve study of Earth's origin, processes, and evolution

An intention of the OP camps is to expose campers with disabilities to a wide range of men and women engaged in science, mathematics, engineering and technology. Credit for creating and directing the OP camps goes to Dr. Richard Radtke, an active research oceanographer who is quadriplegic and unable to move from the neck down. With his faithful service dog at his side, Dr. Radtke relies on headphone-like assistive technologies to operate his computer, use telephones, and amplify his voice. He models for all of us that a person with such severe disability can function and lead an active, professionally productive, and personally fulfilling life. He is surrounded by supportive people who share in his work and well-being, and who, in turn, learn much from him about living life to its fullest. Aptly, he named his project "Ocean of Potentiality" setting out to provide tangible support to youth with disabilities in Hawaii as they envision futures and prepare for careers. Dr. Radtke has drawn together a team of adults with and without disabilities, including scientists, science educators, special educators, parents and other volunteers, to plan and carry out Ocean of Potentiality activities.

According to Dr. Radtke, a youth in a wheelchair is more likely to hear "You could become a counselor who works with others who are disabled" than to be asked "Have you ever thought about doing something in the sciences?" Thus, he envisioned OP camps as an opportunity for youth with disabilities to expand their own visions and explore their own potentials and abilities. As a result of the OP camps, not all the campers will be like my buddy Paul, who suddenly realized that he, too, could aspire to work in the sciences. But, we know that the OP camps have been life-changing events for many campers with disabilities, who experienced things for the first time, saw role models with and without disability and began thinking about new possibilities for themselves, encouraged by adults and each other to expand their own world views. As we implement Understanding 7 in GSL, envisioning career possibilities for our students and ourselves, surely we can make our possibility thinking more inclusive and more expansive, too.

4. MAKING GLOBAL SCIENCE LITERACY FOR ALL A WORLDWIDE REALITY

My purpose in writing this chapter was to advocate that Global Science Literacy can—and should—embrace all learners, including learners with disabilities. I drew from my own experiences with OP science camps to demonstrate how ESE understandings support inclusion of youth with disabilities. Admittedly, these OP camps are unique, costly and time consuming, so I realize that they will not be widely emulated. Nevertheless, wherever the ESE understandings are implemented, they can be carried out in ways that create inclusive, enabling, and accessible learning environments for all.

"All" includes everyone, regardless of age, sex, cultural or ethnic background, social groups, geographic differences, or abilities or disabilities. Because United States federal laws are driving education reforms to implement inclusive teaching practices, educators are transforming their thinking about what is possible for all students, with or without disabilities. As the Global Science Literacy movement seeks to restructure curriculum and implement best educational practices worldwide, GSL now has the opportunity to model effective inclusion of all students, and to make "global science learning for all" a reality.

REFERENCES

American Association for the Advancement of Science. (1993). *Benchmarks for science literacy.* New York: Oxford University Press.

Banks, J. A. (2000). The social construction of difference and the quest for educational equality. In Brandt, R. E. (Ed.), *ASCD Yearbook 2000 Education in a New Era,* (pp. 21-46). Alexandria, VA: Association for Supervision and Curriculum Development.

Council for Exceptional Children. (1999). *IEP Team Guide.* Reston, VA: Council for Exceptional Children (Author).

Choate, J. (2000). *Successful inclusive teaching.* Boston, MA: Allyn and Bacon.

Daniels, S. M. (1990). The meaning of disabilities: Evolving concepts. *Assistive Technology Quarterly,* 1, 3.

Eisenhower Southwest Consortium for the Improvement of Mathematics and Science Teaching. (1996). Putting numbers on performance. *Classroom Compass,* 2, 2.

Individuals with Disability Act (IDEA) of 1990, Publ. L. No. 101-476; Reauthorized in 1997, Publ. L. No. 105-17.

Johnson, D. W., and Johnson, R. (1994). *Learning together and alone: Cooperative, competitive, and individualistic learning.* Englewood Cliffs, NJ: Prentice-Hall.

Kagan, S. (1993). *Cooperative learning.* San Juan Capistrano, CA: Kagan Cooperative Learning.

Kame'enui, E. J., and Simmons, D. C. (1999). *Toward successful inclusion of students with disabilities: The architecture of instruction.* Reston, VA: The Council for Exceptional Children.

Klemm, E. B. and Avery, Q. (1999). Linking reforms: Environmental education and special education. *Proceedings of the 28th Annual Conference of the North American Association for Environmental Education.* Cincinnati, OH.

Klemm, E. B. (1999). Using Earth systems education in field settings for all students, including students with disabilities. *Proceedings of the 1999 Annual Conference of the North American Association for Environmental Education.* Cincinnati, OH.

Klemm, E. B., Radtke, R. L., and Skouge, J. (2000). Inclusion of all students in fully accessible, technology-supported, field-based marine science camps: The ocean of potentiality project. In Clark, I. F., (Ed.), *Proceedings of the 3rd International Conference on Geoscience Education:* Vol. 1 (pp. 71-74). Sydney, Australia: Australian Geological Survey Organisation

Klemm, E. B., Skouge, J. R., Radtke, R. L., and Laszlo, J. R. (2000, Winter). Ocean of potentiality: Fully accessible science camps. *Journal of Science Education for Students with Disabilities,* 8, 22-29.

Lee, O. (1999). Equity implications based on the conceptions of science achievement in major reform documents. *Review of Educational Research,* 69, 1, 83-115.

Mastropieri, M. A., and Scruggs, T. E. (1992). Science for students with disabilities. *Science Education,* 62, 4, 377-411.

National Institute on Disability and Rehabilitation Research. (1998). Notice of proposed long-range plan for fiscal years 1999-2004. Federal Register, 63 (206), 57189-57219.

National Research Council. (1996). *National science education standards.* Washington, DC: National Academy Press.

North American Association for Environmental Education. (1998). *Environmental education materials: Guidelines for excellence.* Troy, OH: Author.

North American Association for Environmental Education (1999). Excellence in environmental education: Guidelines for Learning (K-12). Rock Spring, GA: Author.

Pipho, C. (2000). Governing the American dream of universal public education. In R. S. Brandt, (Ed). *ASCD Yeabook 2000: Education in a New Era* (pp. 5-20). Reston, VA: Association for Supervision and Curriculum Development.

Rutherford, F. J., and A. Ahlgren. (1990). *Science for all Americans.* New York: Oxford University Press.

Scruggs, T. E., and Mastropieri, M. A. (1993). Current approaches to science education: Implications for mainstream instruction of students with disabilities. *Remedial and Special Education,* 14, 11, 15-24.

Seelman, K. D. (1998). Change and challenge: The integration of the new paradigm of disability into research and practice. Paper presented at the National Council on Rehabilitation Conference. Vancouver, WA. Available: http://www.ncddr.org/new/speeches/ncre/index.html.

Siegel-Causey, E., McMorris, C., McGowen, S., and Sands-Buss, S. (1998). In junior high you take earth science. *Teaching Exceptional Children*, 31, 1, 66-72.

U. S. Department of Education. (1998). *20th annual report to Congress on the implementation of IDEA.* Washington, DC: Office of Special Education and Rehabilitation Services.

Vaughn, S., Bos, C. S. and Schumm, J. S. (1997). *Teaching mainstreamed, diverse, and at-risk students in the general education classroom.* Needham Heights, MA: Allyn and Bacon.

Vygotsky, L. S. (1978). *Thought and action.* Cambridge, MA: Massachusetts Institute of Technology.

Williams, B. (1999). Diversity and education for the 21st century. In D. D. Marsh (Ed.), *ASCD Yearbook 1999: Preparing Our Schools for the 21st Century* (pp. 45-64). Alexandria, VA: Association for Supervision and Curriculum Development.

CHAPTER 13: DEVELOPING THE CONCEPT OF DEEP TIME

Roger David Trend
University of Exeter, UK

1. INTRODUCTION

Earth Systems Understanding 5 (Table 1, Chapter One), one of the seven basic science understandings that form the foundation of Global Science Literacy, establishes the concept of 'deep time' as being important in students' realization of many modern concepts in science--especially that of organic evolution. Little as been done, however, in the design of curriculum materials to develop the concept nor in conduction research studies to determine how the concept could be more adequately presented to students. In this chapter, I report on research into the understanding of this concept across UK society. Some of this research is published, some is in press and some is currently in train. The chapter ends with a proposal for a Deep Time Framework of Pivotal Geo-events.

2. DEEP TIME: A CORE ELEMENT OR A CRITICAL BARRIER?

Given the concerns about developing science literacy in UK schools reflected in the *Beyond 2000* report (Millar and Osborne, 1998) and elsewhere, it is proposed that the concept of deep time (McPhee, 1981) deserves special attention, for several reasons. First, it is a key concept across the breadth of science, notably in evolutionary biology, astrophysics and geoscience. Second, it has received almost no research attention in relation to conceptual learning across UK society. Third, it is encountered very often in the popular media yet rarely receives sound treatment beyond the "millions and millions of years" approach. Fourth, evidence indicates that this concept acts as a critical barrier to the development of science literacy (see below). Fifth, Earth events in deep time provide a challenge to the experimental-predictive approach to science, as discussed by Gould (1986).

    Deep time is not entirely ignored in the UK science education literature, although its status has waxed and waned over the decades. In the context of science and religion, Downie and Barron (2000) report on the changing attitudes of Scottish biology undergraduates towards evolution (and four other major theories), noting that up to 11% "rejected the occurrence of biological evolution" (p. 139). They also note that the students perceive the theory of plate tectonics to be based on dubious evidence, although they are very content with the validity of the three other major science theories (dealing with CFCs, cigarettes and acid rain). Both evolution and plate theories have strong deep time implications, although the authors do not address this issue. Perhaps respondents' perceptions and conceptions of deep time

V.J. Mayer (ed.), Global Science Literacy, 187–201.

might be relevant in the explanation.

Millar and Osborne (1998) advocated the wider use of narrative in developing science literacy in relation to deep time. If their curricular proposals are to be implemented, possibly at the 2005 revision of the UK National Curriculum, we need to investigate how people conceive deep time and how appropriate learning can be fostered. The same authors recommended that 14-16 year-old children should engage with the idea that "evidence is often uncertain and does not point conclusively to any single explanation" (p. 5). Donnelly (1999) made a similar point, advocating the encouragement of children's multiple interpretation of scientific evidence, as with historical evidence. With its reliance on evidence which quite definitely "is often uncertain and does not point conclusively to any single explanation", geology in its deep time context provides an appropriate context for teaching in this way.

Deep time is at the heart of geology, as noted by major authors over the decades: Whewell (1837), Kitts (1977), and Gould (1987). During geology's "Heroic Age of Geology" around 1780 to 1830, Porter, (1973), Hutton (1788; 1795) and Toulmin placed geological time center stage (McIntyre, 1963; Toulmin, 1962). These authors were not the first to address the concept, but they were leading figures who emphasised it at a time when geology was rapidly maturing into the leading science of the time. Since then deep time has remained a core concept in geology (Albritton, 1963) and in other fields of science. As such it has a crucial role in global science literacy education.

More recent commentators have drawn attention to the critical nature of the deep time concept. McNeill (2000) examines Twentieth Century environmental change in a deep time context. Gould (1986) contrasts two major styles of science: the experimental-predictive and the historical, asking "how can a naturalist do history in a scientific way, especially given the poor reputation of history as a ground for testable hypothesis? How can history be incorporated into science?" (p. 61). After all, writes Gould, "history is the domain of narrative - unique, unrepeatable, unobservable, large-scale singular events" (p. 61) whereas science "traffics in process" (p. 62) and involves "observation, simplification to tease apart controlling variables, crucial experiment, and prediction with repetition as a test. These classic 'billiard ball' models of simple physical systems grant no uniqueness to time and object..." (p. 64).

Gould (1987) claimed that: "deep time is so difficult to comprehend, so outside our ordinary experience, that it remains a major stumbling block to our understanding" (p. 2). Three years later, in his plea for a "proper scale for our environmental crisis", he argued that deep time is "all too rarely grasped in daily life" (Gould, 1990, p. 24) and that the "measuring rod of a human life" is generally inadequate in helping people to engage with geological periods of time.

Given that deep time is widely accepted as a distinctive high-order concept, can it also constitute a critical barrier to further science learning? Various authors have attached different labels to these high-order concepts: key; critical; fundamental; anchoring; basic; and so forth, but the common strand is that the concept occupies a pivotal place in a hierarchy (Gagne, 1985), either of the discipline or of learner progression. Schoon and Boone (1998) suggest that the holding of significant misconceptions concerning deep time may be a cause of low self-efficacy and, therefore, a fundamental barrier to wider science understanding. Hawkins (1978) refers to "critical barriers" and Clement *et al.* (1989) develop the idea of an "anchoring conception" (abbreviated to "anchor") as "an intuitive knowledge structure that is in rough agreement with accepted (physical) theory" (p. 555). Failure to grasp an "anchor concept" places a constraint on subsequent learning. Novak (1988) also uses the term "anchor", but in relation to concepts which may or may not correspond to the scientific consensus.

Research evidence is lacking so we are not yet approaching a secure answer to the questions posed above. However, the research findings reported below reveal several areas in which perceptions and conceptions of deep time differ tremendously from the scientific consensus. It is argued that deep time is a science concept which has a strong claim to be labeled as a "critical barrier" or a "fundamental barrier" and that the conception/consensus gaps should give cause for concern and lead to appropriate curriculum development. Such work is necessary if deep time is to figure as an essential ingredient of a scientifically literate person.

3. RESEARCH INTO CONCEPTIONS AND PERCEPTIONS OF DEEP TIME

3.1. Background to the Deep Time Research Project

Little research has been undertaken to probe children's understanding of geological time (Dove, 1998; Trend, 1998; 2000; in press; submitted). In the early years of the "alternative frameworks" movement, Ault (1982) examined understandings of deep time among children aged 7 to 11 years. These children applied common sense interpretations to (compost heap) successions but they had difficulty transferring this conceptualizing to rock stratigraphic sequences. Ault found that any lack of knowledge about specific Earth history events was no barrier to grasping time-related concepts such as succession, superposition and correlation. Other authors include direct or indirect reference to deep time alongside other geoscience concepts (e.g. Happs, 1982; Leather, 1986; Marques, 1988; Schoon, 1989; Oversby, 1996; Marques and Thompson, 1997). Ault (1994) provides a review of this research in the context of research on problem solving in Earth science.

The research summarised very briefly in the following paragraphs was undertaken between 1998 and 2000 in order to investigate the nature of deep time conceptions across UK society. The research continues. In all situations the Earth events are labelled as "geo-events". To date the following respondent categories have been studied:

- Children aged 10 and 11 years
- Students aged 17 years
- Trainee (pre-service) Primary Teachers
- Serving (in-service) Primary Teachers

3.2. Understanding among children aged 10 and 11 years

Marques and Thompson (1997) found that children aged 10-11 years fail to distinguish between the formation of the Earth and the Universe. They do not conceive of anything prior to the Earth's existence. Similarly, many children 4 or 5 years older are conscious of something akin to a *Big Bang* but do not separate this event in time from that of the Earth's formation. Trend (1998) reported multi-instrument research into conceptions of deep time among 177 10- and 11-year-old children from 4 UK schools. Card sequencing was identified as a reliable technique for exploring mis- or pre-conceptions around deep time and for yielding quantitative measures of consensus among the children. Three different card-sequencing tasks were used, involving a total of 23 geo-events across three sets of cards.

Among the detailed findings reported, the geo-event, which caused the greatest disagreement among the children over its relative sequential position, was *Big Bang*. Its average rank was second, **after formation of the Sun.** It was suggested that such high disagreement might result if the children fell into two categories: (i) those who have encountered the phrase *Big Bang* and who identify it, however loosely, with extreme age and (possibly) the beginning of the universe, and (ii) those who have never previously encountered the phrase and who therefore responded randomly. It was concluded that the children perceive certain events to have occurred at a very early time (*Extremely Ancient*) and the others more recently (*Less Ancient*). The first are concerned with the genesis of fundamental units: planets, stars and the Universe and the second with changes affecting those bodies. Among the *Less Ancient* events, *The Ice Age* produced the greatest disagreement among these 10- and 11-year-old children, although it was ranked in its correct position. A proposed ancient origin for continents was notable (for its accuracy), although there was poor correlation between the timing of Earth's formation, the first rocks, the first volcanoes and the first life. Beyond that broad two-category grouping lies confusion, pre-conception and opportunities for teachers to foster more secure science learning. The following conclusions were reported:

1. Children at 10 and 11 years of age perceive geological events as falling into two broad groups: the *Extremely Ancient* and the *Less Ancient*.
2. The first rocks and the first volcanoes are not closely associated, temporally, with the formation of the Sun and planet Earth
3. There is considerable collective confusion about the relative timing of the *Ice Age*.

3.3. Understanding among students aged 17 years

This research (Trend, submitted) involved two instruments and two samples of 17-year-old students. The first instrument comprised list writing by 50 students drawn from two UK schools. Respondents were each asked to write down a list of major Earth events that had occurred in the past, since the time of planet Earth formation. Minimal stimulus was provided. The second instrument involved concept mapping by 86 students in an additional six schools. In the first study a total of 775 geo-event statements was recorded across 50 respondents (mean 15.5 per respondent). Results from both studies were presented in relation to the lithosphere, biosphere, hydrosphere and atmosphere.

Lithosphere Sixty two percent of respondents named *Big Bang* as a major geo-event, despite their brief being related to the period *since the formation of planet Earth*. Eighteen percent linked *Big Bang* directly with the formation of planet Earth. Such confusion over the temporal relationships of these two major geo-events matches that reported above for 10- and 11-year-old children and below for trainee teachers and serving teachers. Although "the (continental) assembly process generally takes much longer than fragmentation" (Condie, 1997, p. 29), such fragmentation was seen by many as a major geo-event, but coalescence much less so. Young people have an awareness of continental fragmentation and perceive it in isolation from other major geo-events: they are unable to locate it in any deep time scale. They do not perceive continental coalescence as a major geo-event or process and are, therefore, unaware of any super continent cycle. Respondents made almost no reference to Earth materials. Young people do not perceive such everyday geological materials in temporal terms: they merely have an existence and possess certain (stable) properties. If they have a history, it is perceived to be unimportant.

The concept-mapping instrument generated 120 links between *Big Bang* and other geo-events, 70% being incorrect (by current scientific consensus) and 24% correct (most of these linked *Big Bang* with *Universe*). Links were freely composed by respondents between the concepts provided. Results also show that the introduction of absolute time does nothing to help respondents express their understanding of geo-event sequence through deep time.

Biosphere The research found that selected events in the history of life are widely perceived by 17-year-old people. One half of all named geo-events related to life. The emphasis was on the first appearances of new taxa, followed by evolutionary change rather than on taxa disappearance (almost all in relation to dinosaurs). Extinction was mentioned only by seven respondents and informal phrases such as "wiped out" and "killed off" were used only eight times. Extinctions, either mass or background, do not figure large in respondents' perceptions, despite the high profiles across society of dinosaur extinction and biodiversity.

Hydrosphere and Atmosphere The highest-ranking item for the hydrosphere and atmosphere was Ice Age (presumed by the researcher to be the Quaternary one), ranked at eighth. Other research shows that both pre-service and serving primary teachers have little interest in this geo-event and encounter it relatively infrequently in the classroom (Trend, 2000). Furthermore, as noted above, Ice Age generates the greatest confusion over relative timing among 10- and 11-year-old children (Trend, 1998).

The concept-mapping instrument generated 100 links between *Ice Age* and other geo-events, 39 of them correct, 49 incorrect and 12 strictly correct but suggesting poor understanding. Many identified contemporaneity with Woolly Mammoths. The most-cited incorrect link was with *Dinosaurs*, most indicating that the Ice Age caused dinosaur extinction. This merging of several major geo-events in time has been reported for 10- and 11-year-old children (Trend, 1998) and pre-service teachers (Trend, 2000). People have difficulty in distinguishing between climatic cooling *per se* and the Ice Age, both of which are perceived by respondents as having caused mass extinctions (including both dinosaurs and Woolly Mammoths). Climate change was widely cited by these 17-year-old students and several of them related climate change to extinction and bolide impact. The formation of the atmosphere was cited by thirteen respondents and a further eight referred to its evolution or development, including one who identified "first, second and third atmosphere". Respondents perceived Earth history more in terms of biological and lithological evolution than climatic and atmospheric change, although fluid Earth science matters were not entirely ignored

3.4. Understanding among pre-service primary teachers

This research (Trend, 2000) was undertaken with 179 pre-service primary teachers (broadly 5-11 years, specializing at 7-11 years) and focused on:

- Their interest in geoscience issues and geo-events
- Relationships between their personal interest and routine classroom agendas
- Conceptions of geo-event chronology and the extent to which the provision of a numerical geological timescale influences their representation of that chronology
- Their approach to the enhancement of children's learning in history and in geology with equal effectiveness

Three research instruments were used: questionnaire, card sorting and responding-to-objects. In order to obtain a clear picture of geoscience in primary classrooms, some parallel research was undertaken with 37 serving teachers.

Questionnaire Instrument The questionnaire comprised three main sections, each of 20 items. The first section addressed respondent's personal interest in geoscience and comprised 20 geoscience topics and events, such as "earthquakes", plate tectonics", "rocks, local or otherwise" and "dinosaurs and their extinction". The bottom end of the 5-point response scale was "Personally I dislike this topic; I have no interest in it whatsoever; I have no enthusiasm for it; I have no wish to learn more about it "but, of course, I would if it were essential for the curriculum". The top end was "Personally, I enjoy this topic very much; I have great interest in it; I have great enthusiasm for it; I am keen to learn more about it". The second section of the questionnaire was almost identical to the first, but focused on the extent to which respondents encountered the 20 topics in the classroom.

The 179 respondents made full use of the five-point response scale in describing their interest in 20 selected geoscience issues and geo-events. These ranged from highest for *Origin of Solar System* to lowest for *Minerals and Crystals*. By contrast, the lower end of the scale dominated in responses on *Classroom Encounters* of such material. The lowest was for *Ice Ages* and the highest for *Origin of Solar System*. Three of the 20 items generated high scores in both *Interest* and *Encounters*:

- Origin Solar System
- Earthquakes
- Volcanoes and/or volcanic eruptions

It was proposed that these three items represented the geoscience which pervades society and with which pre-service teachers feel relatively secure. Two items generated low scores in both *Interest* and *Encounters*:

- Ice Age/s
- Development of mountain chains

It was proposed that these two items represented the geoscience collectively avoided by society and with which pre-service teachers feel insecure. They have no immediate links with the National Curriculum and other research (Trend, 1998) shows that there is no consensus among 10-year-old children concerning the relative age of *Ice Ages*. Five items were relatively popular among pre-service teachers but serving teachers reported low occurrence in classrooms:

- The origin or formation of planet Earth
- Big Bang
- Evolution of humans. The first humans
- Plate Tectonics / Continental Drift
- History of geological ideas

It was suggested that these items are representative of the central concepts of geoscience that are deemed too complex or controversial to be introduced in the classroom for 7-11 year-olds. Four further items were relatively unpopular among pre-service teachers but serving teachers reported high occurrence in classrooms:

- Rocks: local or otherwise
- Minerals and/or crystals
- Fossils in general: local or otherwise
- Current landforms and processes: e.g. rivers, cliffs, hills, valleys, plains

The third section of the questionnaire comprised 20 geo-events, which respondents had to allocate to one of nine time period categories, ranging from "less than one thousand years ago" to "more than a million million years ago". This task required respondents to engage with absolute times, although results were analyzed in both absolute and relative terms. Results are discussed below.

Card-Sorting Instrument. Respondents were provided with 21 cards, being the same geo-events used in section three of the questionnaire, with the addition of "present day" to assist physical sorting. Respondents had to arrange (and record) the cards in chronological sequence, according to their perceptions. Card sorting and questionnaire revealed a three-fold structure for perceptions of geo-events, similar to the two-fold structure reported above for 10- and 11-year-old children. The pre-service teachers conceived the 20 geo-events as falling into three distinct time categories: Extremely Ancient (e.g. first rocks), Less Ancient (e.g. Atlantic opening) and Geologically Recent (e.g. first humans). Within the Less Ancient cluster, the timing of Ice Age/s generated the greatest disagreement, as it did for 10- and 11-year-old children (Trend, 1998). It was suggested that the term Ice Age carried no clear temporal or evolutionary connotations, unlike geo-events involving organisms more explicitly.

The questionnaire and the card-sorting instruments required respondents to engage with both relative and absolute time. Results suggested that the role of large numbers in penetrating people's grasp of deep time is of highly questionable value. The provision of real dates (absolute timescale) tended to reduce response accuracy: relative time is more effective in fostering understanding of earth history.

Responding to Objects Instrument. Respondents were seated around a central table on which were placed three historical artifacts and three geological specimens. They were asked to imagine that a child had brought each object into the classroom and they were to record all the possible responses they might make to foster that child's learning. Each object was to be treated separately and each respondent was to work alone. These pre-service teachers demonstrated no strong disposition to establish a link between the present with the past or to select time as a particular stimulus for children's learning.

3.5. Understanding among serving primary teachers

This research (Trend, in press) focused on serving primary (5-11 years) teachers and was similar in design to that reported above for pre-service teachers. The geoscience interests of serving primary teachers in general were found to be relatively uniform whereas there was more variation in classroom encounters of geoscience. The author reported a healthy tendency for teachers to be interested in the geoscience topics actually encountered in the classroom and to be less interested in matters not commonly encountered.

As with pre-service teachers, it was found that the presentation of an absolute time scale led to the demonstration of some extreme misconceptions concerning both absolute and relative dates. They did not perceive the *Formation of the Sun* and the *Big Bang* as separated in time. Indeed, according to respondents' absolute time categories, the Sun was conceived as having been formed before the *Big Bang*. These findings have their parallel with those of Marques and Thompson (1997) concerning children's misconceptions in deep time, with about half of their samples of 10-11 and 14-15 year-olds believing "that the origin of the Earth and life occurred simultaneously" (p. 40). Later they address briefly the matter of scale in time: perhaps the children "felt that these two events occurred a long time ago but they may have felt difficulties in relating the two events chronologically on a suitably large time-scale" (p. 40). This observation corresponds closely with the proposition made elsewhere (Trend, 1998, 2000) and developed further below in relation to the conception of deep time as broad categories within which relative sequence is opaque.

In line with similar findings reported above and elsewhere for young children and pre-service teachers, *Ice Age* generated much disagreement among serving teachers. Also similar to findings with other response categories, it was reported that serving teachers conceive geo-events in deep time as having occurred in three broad categories of time: *Extremely Ancient, Moderately Ancient* and *Less Ancient*. In fact, a more refined but less reliable 5-category model was proposed.

The "responding-to-objects" instrument yielded data on serving teachers' perceptions of history and geology. They were found to be far more comfortable using history artifacts to compare the present with the past than they were using the geology specimens. They placed no special focus on deep time with the geology specimens. They demonstrated neither a disposition to establish a link between the present and the past (strongly history-orientated, 95%) nor any tendency to select time as a particular stimulus for children's learning.

3.6. Common strands from the four research projects

Several **common strands** can be discerned running through the four research projects described very briefly above, although not all evidence for these strands is fully covered in those summaries. For each of the strands described below, it is proposed that an appropriate Deep Time Framework of Pivotal Geo-events would assist the development of science literacy. Indeed, current debates on the nature and significance of science literacy would be enriched by the systematic inclusion of such Deep Time Frameworks. As Mayer (1995, 1996, 1997a, 1997b, 1998) and Mayer and Armstrong (1990) point out, one of the (seven) all-pervasive "understandings" necessary for a sound Global Science Literacy is that "Earth is more than 4 billion years old and its subsystems are continually evolving" (see Table 1, Chapter One).

First, all categories of learner conceive geo-events as occurring in a small number of discrete time categories which may be labeled along the lines of: *Extremely Ancient; Moderately Ancient; Less Ancient*; and *Geologically Recent*.

Second, understanding of various Earth events or processes and their temporal framework are best achieved together. This is particularly important for traditional school-based education. If the teacher lacks the knowledge, understanding and/or confidence in either or both of these aspects, it is suggested that s/he is less likely to respond with powerful learning pointers when study of such an Earth event arises in the classroom curriculum or emerges in general classroom discussion. Understanding of each Earth event should accompany an appreciation of its deep time context. The two should proceed together. However, it becomes a circular matter when the teacher does not possess at the outset a temporal framework for the chronology of geological events. It is suggested that such a teacher is more likely to minimize the impact of a potential (geological) learning opportunity rather than to maximize it. Thus, for example, a mass-media news item about some newly-discovered fossils, a proposed new date for the age of the Earth or a newly-discovered catastrophic event are unlikely to be used effectively by the teacher if s/he lacks a secure geological chronology, since questions about deep time are very likely to be central to any open class discussion of such matters.

Third, there is much geoscience that is currently in people's general awareness, although such conceptions are loosely held, understanding is insecure and a temporal framework is largely lacking. Seventeen-year-old students are able to list many important geoevents and serving teachers revealed a close match between their interests and classroom encounters.

Fourth, the introduction of an absolute time scale reduces the accuracy of people's representation of geo-events through deep time.

Fifth, pre-service and serving primary teachers are more secure in their handling of time concepts in the context of history, compared with geology. There are many reasons for this, including the difference in time-spans involved, but a Deep Time Framework would provide a conceptual structure for their own developing science literacy and, therefore, that of their pupils.

4. A DEEP TIME FRAMEWORK OF PIVOTAL GEO-EVENTS

4.1. Principles

It is proposed that an optimum Deep Time Framework (DTF) of Pivotal Geo-events (See Table 1) should be developed for each category of learner in order to provide a secure foundation for the development of science literacy. Each geo-event is "pivotal" in relation to the learning process, not in relation to the history of planet Earth itself. Each DTF should comprise a **small number** of carefully-selected geo-events, taught in ways appropriate to the learners but always with a strong emphasis on absolute and relative dates and durations: the deep time component.

The DTF is conceived here as a personal construct. It comprises the chronology of key geo-events, together with their absolute dates, which the learner brings to bear when encountering a new geo-event or geoscience phenomenon. For some people this framework is detailed, sophisticated and well developed. For others it is scant, as outlined above. The culmination of this chapter is the proposal for the development of Deep Time Frameworks of Pivotal Geo-events. In that way the science literacy of UK citizens may be more securely enhanced than it has been in recent decades.

4.2. Criteria for Developing Deep Time Frameworks

Each DTF of Pivotal Geo-events should be developed so that it:

1. accommodates and exploits learners' existing interests and encounters;
2. challenges existing conceptions concerning the chronology of deep time events;
3. covers the full history of planet Earth, 4.6 billion years;
4. includes major events in the history of life, such as first and last appearances and mass extinctions;
5. includes major events in continental fragmentation and coalescence;
6. includes events which facilitate links between past and present;
7. addresses some of the main geological materials of planet Earth such as rocks and minerals;
8. includes references to climate change; and,
9. includes small reference to absolute time.

Each major geo-event combines time and place dimensions with rapid or gradual change. Each can be regarded as being equivalent to a high-order concept. For example, understanding the concept of "mass extinction" requires the conceptualization of a host of lower order concepts (e.g., organism, death, evolutionary change, proportion, fossilization) and, for example, engaging with the particular geo-event known as the end-Permian mass extinction requires that conceptualization to have a location in time (and place). Science literacy for many people would include the ability to engage with that major geo-event so that it enriches their Deep Time Frameworks, thereby providing a more secure basis for subsequent learning.

A DTF construct is a type of conceptual framework, with each major geo-event treated as a concept, whether it is part of a cycle or an arrow of time. The linkages made by learners between the dates of geo-events (such as "1,500 million years before" or "around the same time") and the events themselves (e.g. a mass extinction, glaciation or continental fragmentation) would serve to reinforce the DTF and thereby accommodate subsequent learning.

4.3. Exemplar DTF

A DTF for beginning teachers has been proposed elsewhere (Trend, in press). That model takes into account relevant research and the perceived needs of UK teachers. The following brief DTF is proposed for English children aged about 14 years, most of who are moving from Key Stage 3 to Key Stage 4 of the English National Curriculum. The following curriculum content satisfies the criteria given above and takes into account the National Curriculum requirements in Science and Geography (Department for Education and Employment and Qualifications and Curriculum Authority 1999a; 1999b). Taken together, the deep time content items in Table 1 and expanded on in Table 2 constitute a proposed DTF which would be appropriate for UK students towards the end of their (compulsory) secondary school careers, at about the age of 15 years. Each item:

- Arises from the research summarised above, either directly or indirectly.
- Relates to the 9 Criteria suggested above.
- Is in the spirit of Huxley's concentric approach.
- Relates directly to ESS Understandings 4, 5 and 6 (see Table 1, Chapter One).
- Contributes to broad science literacy in terms of knowledge of Earth's evolution.

Table 1. Pivotal geo-events for UK children aged about 14 years

Geo-event	Criteria (see text)									National Curriculum			
										Geog	Science		
	1	2	3	4	5	6	7	8	9	KS3	KS3	KS4S	KS4D
Planetary and Lithospheric													
1. Big Bang; Universe	x	x							x			x	x
2. Formation of planet Earth; Solar System; Sun	x	x	x						x			x	x
3. First crust, early continents			x		x				x				
4. Continental fragmentation	x		x		x	x	x		x	x			x
5. Continental coalescence		x	x		x		x		x	x			x
6. Rocks as environmental records			x			x	x		x		x		x
7. Bolide impact	x	x	x	x				x	x			x	x
8. Rock formation; local examples	x		x		x	x			x		x		x
9. Volcanic eruption	x		x		x	x	x	x	x	x			x
10. Earthquake	x		x		x	x			x	x			x
11. Local lithospheric geo-event	x		x		x	x	x		x		x		x
12. Orogeny		x	x		x	x							x
Biospheric													
13. Dinosaur extinction; end-Cret mass extinction	x		x	x					x			x	x
14. First appearance of modern humans	x		x	x		x			x			x	x
15. First trees; Carboniferous coal forests	x		x	x		x	x		x		x	x	x
16. First mammals	x		x	x		x			x			x	x
17. Cambrian Explosion		x	x	x					x			x	x

Atmospheric and Hydrospheric											
18. Atmosphere origins: change; photosynthesis		x	x					x				x
19. Ice Age	x	x	x			x		x	x	x		x
20. Post-glacial climate change	x	x	x			x		x	x	x		x

Table 2. A deep time framework of pivitol geo-events.

Geo-event	Date MYA	Notes
10. Earthquake	0.0001	Example San Francisco
20. Post-glacial climate change	0.001	10,000
9. Volcanic eruption	0.36	Example Santorini
14. First appearance of modern humans	1	
19. Ice Age	2	Example Quaternary Ice Age
7. Bolide impact	65	Chicxulub
13. Dinosaur extinction; end-Cret mass extinction	65	
16. First mammals	210	
6. Rocks as environmental records	240	Example Scythian Pebble Beds, East Devon, UK
11. Local lithospheric geo-event	300	Example Dartmoor emplaced
8. Rock formation; local examples	380	Example Givetian limestones, South Devon, UK
5. Continental coalescence	400	Example Pangaea formation, 450 – 320 MYA
12. Orogeny	400	Example Caledonian
15. First trees; Carboniferous coal forests	420	Also Westphalian coal forests, 300 MYA
17. Cambrian Explosion	530	
4. Continental fragmentation	680	Example Rodinia fragmentation, 750 – 600 MYA
18. Atmosphere origins: change; photosynthesis	2,400	Start of O2 in atmosphere and surface waters. Also 1,800 MYA as free oxygen pervades ocean/atmosphere system
3. First crust, early continents	4,300	Between c4,600 and 3,800 MYA
2. Formation of planet Earth; Solar System; Sun	4,600	
1. Big Bang; Universe	14,000	Between c12,000 and 15,000 MYA

Some of the geo-events are stated unambiguously, e.g. Big Bang. Others require interpretation and exemplification by the teacher. For example, teachers will need to select rocks (6 and 8) and geo-events (11) that are specific to their own localities, ensuring that Criterion 1 is addressed. No indication is given here on possible learning activities, learning materials, teaching strategies and other matters of pedagogy. No elaboration of content is provided. It is suggested that the teacher should teach the following DTF in such a way as to emphasize the relative timing of the major geo-events, paying particular attention to areas, identified above, which have been shown to cause confusion among learners, for example Big Bang and the formation of the Solar System.

ACKNOWLEDGEMENTS

The author acknowledges with deep gratitude the detailed advice offered by David Thompson in the preparation of this chapter.

REFERENCES

Albritton, Claude C. (Ed.). (1963). *The fabric of geology.* Stanford: Freeman, Cooper and Co.

Ault, Charles R. Jr. (1982). Time in geological explanations as perceived by elementary-school students. *Journal of Geological Education,* 30, 5, 304-309.

Ault, Charles R. Jr. (1994). Research on problem solving: Earth science. In Dorothy L. Gabel (ed.) *Handbook of research on science teaching and learning: A project of the National Science Teachers' Association,* (269-283). New York: Macmillan.

Clement, J., Brown, D. and Zietsman, A. (1989). Not all preconceptions are misconceptions: finding "anchoring conceptions" for grounding instruction on students' intuitions. *International Journal of Science Education,* 11, 5, 554-566.

Condie, Kent C. (1997). *Plate tectonics and crustal evolution.* Oxford: Butterworth-Heinemann.

Department for Education and Employment and Qualifications and Curriculum Authority (1999a). The National Curriculum for England: Science. London: HMSO.

Department for Education and Employment and Qualifications and Curriculum Authority (1999b). The National Curriculum for England: Geography. London: HMSO.

Donnelly, James. (1999). Interpreting differences: the educational aims of teachers of science and history, and their implications. *Journal of Curriculum Studies,* 31, 1, 17-41.

Dove, J. E. (1998). Students' alternative conceptions in Earth science: a review of research and implicatioins for teaching and learning. *Research Papers in Education,* 13, 2, 183-201.

Downie, J. R. and Barron, N. J. (2000). Evolution and religion: attitudes of Scottish first year biology and medical students to the teaching of evolutionary biology. *Journal of Biological Education,* 34, 3, 139-146

Gagne, R. M. (1985). *The conditions of learning and the theory of instruction (4th Edition).* New York: Holt, Rinehart and Winston.

Gould, S J. (1986). Evolution and the triumph of homology, or why history matters. *American Scientist,* 74, 60-69.

Gould, S J. (1987). *Time's arrow, time's cycle.* Harmondsworth: Penguin.

Gould, S J. (1990a). The golden rule: a proper scale for our environmental crisis. *Natural History,* 9, 90, 24-30.

Happs, J. C. (1982). Some aspects of student understanding of two New Zealand landforms. *New Zealand Science Teacher,* 32, 4-12.

Hawkins, David. (1978). Critical barriers to science learning. *Outlook,* 29, 3-23.

Hutton, James. (1788). Theory of the Earth. *Transactions of the Royal Society of Edinburgh,* 1, 209-305.

Hutton, James. (1795). *Theory of the Earth with proofs and illustrations.* Edinburgh: William Creech.

Kitts, D B. (1977). *The structure of geology.* Dallas: SMV Press.

Leather, A. D. (1986). *An Investigation Into the Views Held by 11-17 year-olds of 2 Earth Science Phenomena: Earthquakes and Oil.* Unpublished MSc (Ed) dissertation, University of Keele, UK.

Marques, L. F. (1988). *Alternative Frameworks of Urban Portuguese Pupils aged 10-11 and 14/15 With Respect to Earth, Life and Volcanoes.* Unpublished MA thesis, University of Keele, Keele, UK.

Marques, L. F. and Thompson, D.B. (1997). Portuguese students' understanding at ages 10-11 and 14-15 of the origin and nature of the Earth and the development of life. *Research in Science and Technological Education,* 15, 1, 29-51.

Mayer, V. J. (1995). Using the Earth System for integrating the science curriculum. *Science Education,* 79, 4, 375-391.

Mayer, V J. (1996). The future of geosciences in the pre-college curriculum. In: DAV Stow and GJH McCall (Eds), *Geoscience education and training in schools and universities, for industry and public awareness.* Joint Special Publication of the Commission on Geoscience Education and Training of the International Union of Geological Sciences and the Association of Geoscientists for International Development, AGID Special Publication Series, 19, 183-196. Rotterdam: AA Balkema.

Mayer, V J. (1997a). Science literacy in the global era. *Hyogo University of Teacher Education Journal*, 17, 3, 75-89.

Mayer, V J. (1997b). Global Science Literacy: an Earth System view. *Journal of Research in Science Teaching*, 34, 2, 101-105.

Mayer, V. J. (1998). World War Two, Japan, the United States and global education. *Journal of School Education*, 10, 105-116.

Mayer, V J. and Armstrong, R.E. (1990). What every 17-year-old should know about planet Earth: the report of a conference of educators and geoscientists. *Science Education*, 74, 2, 155-165.

McIntyre, D. B. (1963). James Hutton and the philosophy of geology. In C.C. Albritton (ed), *The fabric of geology*, pp 1-11. Stanford: Freeman, Cooper and Co.

McNeill, John. (2000). *Something new under the sun: An environmental history of the twentieth century.* London: Allen Lane.

McPhee, John. (1981). *Basin and range.* New York: Farrer, Straus and Giroux.

Millar, Robin and Osborne, Jonathan (eds). (1998). *Beyond 2000: Science education for the future.* London: Kings College, London University.

Novak, Joseph D. (1988). Learning science and the science of learning. *Studies in Science Education*, 15, 77-101.

Oversby, John. (1996). Knowledge of Earth science and the potential for its development. *School Science Review*, 78, 283, 91-97.

Porter, Roy (1973). The industrial revolution and the rise of the science of geology. In M. Teich and R. Young (eds). *Changing perspectives in the history of science* (320-343). London: Heinemann.

Schoon, Kenneth J. (1989). *Misconceptions in the Earth sciences: A cross-age study.* Paper presented at the 62nd annual meeting of the National Association for Research in Science Teaching, San Francisco, March/April.

Schoon, Kenneth J and Boone, William J. (1998). Self-efficacy and alternative conceptions of science of preservice elementary teachers. *Science Education* 82, 5, 553-568.

Toulmin, Stephen (1962). Historical inference in science: geology as a model for cosmology. *The Monist*, 16, 142-158.

Trend, Roger D. (1998). An investigation into understanding of geological time among 10- and 11-year-old children. *International Journal of Science Education*, 20, 8, 973-988.

Trend, Roger D. (2000). Conceptions of geological time among primary teacher trainees with reference to their engagement with geoscience, history and science. *International Journal of Science Education*, 22, 5, 539-555.

Trend, Roger D. (submitted). An investigation into the understanding of geological time among 17-year-old students, with implications for teachers' subject matter knowledge. *International Research in Geographical and Environmental Education.*

Trend, Roger D. (in press). Deep time framework: a preliminary study of UK primary teachers' conceptions of geological time and perceptions of geoscience. *Journal of Research in Science Teaching.*

Whewell, W. R. (1837). *History of the inductive sciences from the earliest to the present times.* Volume 3. London: Parker.

CHAPTER 14: HOW A JAPANESE SCIENCE TEACHER INTEGRATES FIELD ACTIVITIES INTO HIS CURRICULUM

Masakazu Goto
National Institute for Educational Policy Research of JAPAN
(Formerly a science teacher in the Miura City Schools)

1. INTRODUCTION

Global Science Literacy (GSL) as a curriculum construct, seeks to broaden students understanding of the nature of science. It is based on Earth Systems Education (ESE). Its basic thesis is that science is the process that we as humans use to understand the world we live in and its environment in space. Therefore, all science instruction should start with some aspect of the Earth systems (biosphere, solid earth, atmosphere and hydrosphere), and expand to the solar system, or the universe. As we live on Earth, the central and important subject for science teaching should be the Earth, particularly our familiar natural environments. Wherever possible we should start with fieldwork in familiar environments and expand our study from local natural environments to such wider areas as regional environments, national environments, global environments and last, the universe. GSL and ESE focus on the science knowledge that will enable the world's citizens to understand the need for global efforts at environmentally sustainable economic and social development. Outdoor education and fieldwork provide an important basis for learning about the Earth systems. Therefore, the fieldwork program will be central to efforts to accomplish the goals of science education of the future.

2. GENERAL DESCRIPTION OF THE INTEGRATED CURRICULA

As a science teacher in Minami-shaitaura lower secondary school, one of the lower secondary schools of Miura City, I developed curricula for the seventh and ninth grades of my school that addressed integrated learning centered on fieldwork (outdoor learning). Because my students' fieldwork was in their local area, it provided practical integrated learning closely related to their daily lives. The collaborative study among students in fieldwork not only teaches them how to cooperate in scientific research but also facilitates their communication abilities.

Every student did five field related projects, each with a different purpose. The project content was organized to help the student to understand the local environment and to acquire the knowledge and skills needed to conduct the fieldwork through a range from introductory level to advanced level. On the last field project, each student developed a proposal for his or her fieldwork that was

V.J. Mayer (ed.), Global Science Literacy, 203–216.
© *2002 Kluwer Academic Publishers. Printed in the Netherlands.*

subject to my advice and approval. The students then conducted their study on their own time, after school and on weekends. Therefore, they had ample time to complete their research. In their research, they were often required to know content from subjects other than science. In this way, their interests were broadened from science to other subjects. The integrated learning I organized included not only science but also homemaking, Japanese languages, fine art, social study, technology (its content related with science), mathematics (graphs and statistics), and English. I collaborated with several teachers of other subjects in order to coordinate and facilitate students' learning goals established in these subjects. As a result, students deepened their understanding of science and became more interested in science through seeing the relationship between science and other subjects.

Integral to my curriculum is the learning network. To improve my curricula I developed a plan that made use of such community resources as museums, universities and institutes and their personnel. For example, I invited specialists and local experts on bird watching, geology, and flora and fauna of the Miura Peninsula to visit my classes and instruct my students about their interests. I also took interested students to the museum for their study on Saturday afternoons and Sunday. After students had completed their research, I arranged to exhibit their results in the local community hall so that they could communicate them to local citizens. The learning network has many levels. For example, the network of school subjects, the network of students' communication, the network of school and community facilities.

My school is now linked to the Internet so in the future students will also be able to communicate their research to students in other schools and people out of school. This capability will further improve their in-school learning. Students will extend their learning from the school to their local community, from their school to the nation of Japan, and finally from their school to the world through the Internet. They will make use of various information sharing networks and thus **expand** their learning **organically**. I call this method of fieldwork-centered and student-centered science learning, Expansive and Organic Learning (see endnote). It is based on the basic principles of Global Science Literacy.

Considering the primary purpose of school education, teachers should value more highly improving students' various abilities in addition to the ability to answer questions and solve problems. Teachers must help students foster the ability, interest, attitude, and skill to continue to learn for the enrichment of their lives during and after their school years. Integrated curricula and student-centered instruction are good examples of instructional processes that can increase student interest in science. Learners, whether they are students or teachers, expand their worldviews through learning. They find out about themselves and how to live and enjoy their lives. This contributes to a democracy that is supported and can only be sustained by socially and scientifically literate people.

2.1. Advantages for a Field Work Centered Curriculum

I have found many advantages in a fieldwork-centered curriculum. Each of those I list below are ones that I have found valid through my use of the curriculum and the learning procedures that I have described.

- Students like fieldwork.
- Students can develop their inquiry skills according to their ability level during fieldwork.
- Many people (adults) have hobbies related to nature such as photography and writing. Children and adults can share their hobbies. Therefore, fieldwork relates to lifelong learning.
- Fieldwork can be related to many subjects. Therefore, the network among school subjects can be easily established.
- Students can easily be assisted in their study by specialists from the local museum and other community resources.
- Fieldwork is easily related to students' daily life and can be developed with the collaboration of local people, especially their parents.
- Students can access many networks while completing their field research.
- Students investigate flora and fauna, weather and geological phenomena in their local area and can relate what they have learned to similar phenomena elsewhere in Japan and the world.
- Students' real experiences are more effective in learning than simply reading books and examining pictures.

2.2. Teaching Method

The teaching methods I used were inquiry-based, student-centered and interactive approaches and included team-teaching. I also included several kinds of fieldwork for different purposes, including both teacher directed and student directed fieldwork. I also used individual to small and large group activities. As students developed and expanded their learning from science to other subjects in these curricula, I cooperated with teachers of other subjects for curriculum planning. When students made use of community resources in their learning, I coordinated the teaching with the specialists.

2.3. Teacher's Role

In the fieldwork-centered curricula, science teachers should correlate and integrate the classroom science lesson with the fieldwork. They should also show students how to apply their knowledge and skills mastered in the classroom to the natural world. Therefore, they should carefully plan the daily classroom lesson as well as the fieldwork. In addition, science teachers should play a central leading role in coordinating and organizing the fieldwork-centered curricula with other subject teachers because science is focused as a central subject in these curricula. Finally, they should coordinate students' questions of the community specialists related to their discoveries and participate in the informal learning experiences that occur outside of the school.

3. THE SEVENTH GRADE CURRICULUM - THE BIOSPHERE

3.1. The Character of the Curriculum:

The seventh grade students begin learning science just after starting lower secondary school. During this introductory period, it is very important to create an interest in science. I emphasize hands-on and direct experience, and life-related science integrated with other subjects. Therefore, I organized the field-centered biological curriculum in which students could learn to use the encyclopedia on flora and fauna and the fieldwork kits, how to plan and conduct research on the local flora and fauna, and how to engage in inquiry-based learning in cooperation with others. I also designed five field projects ranging in locale from schoolyard to city area and from teacher assigned topic to student selected topic. Seventh grade students learned how to investigate by watching TV programs on nature observation and developed their skills through the five field projects.

The fieldwork-centered integrated learning in this curriculum included not only science but also homemaking activities, such as cooking plants collected during the fieldwork; Japanese language instruction such as writing a scientific essay and reading stories related to flora and fauna; fine art such as sketching plants and carving a plate from wood; social study including local human geography and history; and English since the students often communicate in English during their field work. Students utilized the Yokosuka city museum and the Kanagawa prefectural museum, called The Life Planet Earth Museum, and The Nature Conservation Society Museum in Miura for their study.

3.2. Characteristics of the Five Student Field Projects

First fieldwork project: Collaborative study, in small groups of six students each, on the kinds of flora and fauna in the schoolyard:
Six students in a group investigated kinds of plants and their environments (light or shadow, humidity and hardness of soil, etc.). They learned how to make use of the plant encyclopedia and how to investigate cooperatively and collaboratively in a group before the fieldwork. After their investigation and summary of their study, they had a presentation before other students. Students learned how they made the scientific study such as what the science fieldwork investigation was, how they investigated in a group, how they summarized and how they presented.

Second fieldwork project: Individual investigation and collaborative summary of the distribution of three kinds of dandelions in the school district:
Students had homework designed to apply their knowledge and skills to their investigation of the distribution of three kinds of dandelions near their houses. They collected information on their locations in the school district and mapped their distribution, which helped them understand more about their local environment.

Third fieldwork project: Collaborative investigation, including all students, on the distribution of three kinds of dandelions in their city:

This project expanded the dandelion distribution study from the school district to the entire city. The dandelion study was combined with the exercise of orienteering skills, which assisted them in mapping the plants. They developed a map of the distribution of dandelions for the city.

Fourth fieldwork project: Individual investigation on a topic selected by the student:

This is the longest project. They spent three or four school hours for investigating their own idea and task according to their interest in flora and fauna. Almost all students investigated their school forest. They made use of the special investigative apparatus and kits that I had developed. Students developed a deeper understood of their local natural environment through this project.

5th fieldwork project: Individual activity to classify flora and fauna and make up floral specimens:

This is the final project. Students applied their knowledge and skills on how to classify flora in the schoolyard. They collected at least ten kinds of local flora, identified them and made display specimens.

3.3. Relationships with Other Subjects

Japanese (Language): Students learned from an essay about the wild birds in Japan during their Japanese class. In cooperation with the Japanese teacher, science teachers took their students to observe and investigate the birds near the school. A local expert on bird watching was invited as an instructor. They appreciated bird watching guided by him much more after knowing about the birds through reading the essay in the Japanese lesson.

Homemaking: Students investigated flora and fauna in the fieldwork of the science lesson. After that, with assistance of the homemaking teachers, they cooked and ate some of the plants and animals such as fried vegetables with cone flour, grilled horsetail with sweet soy sauce flavor, dandelion coffee and fried fish and beetles. That experience made students familiar with the local environment.

English: Students wanted to make use of English and communicate with each other through English. The fieldwork was a nice opportunity for them to use English by interacting with real materials (flora and fauna) and each other in the outdoors. They enjoyed using English in their science lessons and became proud of their communication abilities in English. Even the students who did not like the regular English lessons showed their interest in English communication. English teachers developed some English textbooks like *Encyclopedia on the Miura Peninsula* and *Flora and Fauna on the Muira Peninsula* in cooperation with science teachers. Science and English teacher-teams made the study of science and English interesting through experiments and fieldwork conducted in English. Eighth grade students also made paintings of some plants and wrote a poem about them in English.

Fine art: Seventh grade students drew a sketch of a natural object found in the outdoors. Eighth grade students also carved a wood plate whose motif was the flora and fauna found in nature. Seventh grade students' knowledge, understanding and experience on nature in the science fieldwork were helpful for drawing their sketch. Eighth grade students could also select the flora and fauna as a motif with some knowledge about nature, in making their sculpture work. Science teachers helped fine art teachers to hold these lessons in the fine art class.

Social studies: Social studies are related to the fieldwork because students learned about their local community through investigating it. They became interested in the local history and the environments where plants grew. They also became acquainted with environmental issues such as nature conservation, contamination, and garbage problems. The fieldwork deepened the relationship between science and social study. In the third field project students collected many beverage cans thus cleaning up and beautifying their local area.

Other subjects: In the future such integrated learning can be expanded by including activities such as those related to music and technology (making musical instruments and playing them in a natural cave), and physical education (country trekking while appreciating the beauty of nature).

3.4. The Learning Network

Making use of public facilities and personnel: Students found many plants and animals, and identified different geographical features in their local areas. Teachers could not respond to all of their findings and ideas. Therefore, the students needed to make use of local specialists, such as museum curators, who could support and facilitate their learning. Therefore, teachers established a learning network including school and out-of-school resources. Teachers not only taught science in the class but also coordinated such a learning network.

The museum: Students found many plants and animals. Subsequently, some students visited the museum on Saturday afternoons or on Sundays so that curators might help identify their specimens. They had an opportunity to see how specialists identified and investigated flora and fauna. Some students interested in nature attended Sunday field excursions conducted by the museum staff.

The nature conservation society: The nature conservation society offers a field excursion once a month. Teachers often advised students interested in nature to join these trips. Some students joined such field excursions as bird watching and nature walks. Science club students often participated in them during the three years they attended my school. They reported on their activities in club meetings and participated in science competitions. In 1996, several science club members were awarded the Governor's Prize for their excellent work

Development of teaching materials: Teachers developed the following teaching materials to support and facilitate students' study and investigation of the local nature.

1.   *Our Native Place, Miura*
2.   *Plants Encyclopedia on The Miura Peninsula*
3.   *Flora on the Miura Peninsula*
4.   *Introduction to Nature Observation*

4. NINTH GRADE CURRICULUM - THE SOLID EARTH

4.1. The Character of this Curriculum

The ninth grade students had obtained knowledge and skills for fieldwork through intensive fieldwork experiences in seventh and eighth grades. They also took their safety and health into account during these field activities and became accustomed to working cooperatively and collaboratively in inquiry-based fieldwork.

Ninth grade science teachers organized five fieldwork projects with different purposes. The first field project was a student-initiated task on geological aspects of the local environment. It was assigned so that students would become more interested in the classroom and lab lessons. They identified interesting findings on local geology in this field project, recorded them, and sketched the outcrop. They were then able to relate their field data to the classroom science lesson. They paid more attention to the science lesson and asked questions about their findings. The 9th grade students exchanged their papers once a week with science teachers. The science teachers commented and answered students' questions on their findings.

Science teachers organized the other four field projects in this curriculum so as to improve students' understanding of the local geology. In the second field project, students were asked to trace a key strata and a famous active fault near the schoolyard. Some students were able to trace the key bed and the fault even though this was a difficult problem. In the their third field project students collected samples of rocks and minerals from the local strata. In the fourth fieldwork project students applied their knowledge and skills to the local nature on Saturday afternoons. Homeroom teachers, as well as other teachers, assisted science teachers to guide their students on a field trip to Kenzaki, the southern tip of Miura Peninsula. They all participated in field activities with students and shared information with them. Science teachers developed a color-illustrated, 200 page book titled *Field guide on the geology in Miura City*. This book was given free to all ninth grade students in Miura City by the Board of Education.

The ninth grade students performed the last fieldwork task as a summer vacation assignment using the field guide as a reference. Their assignments resulted in papers on the geology of the peninsula. Some of them were more than 100 pages long and most included sketches, paintings and photographs as well as graphs and figures. Students applied a variety of skills such as graphing, design of fine art and sentence composition in Japanese in developing their papers. Science teachers developed the New Standard of Evaluation, a rubric for evaluating students' products. This standard included the assessment of scientific ideas and thinking, precise knowledge, skills and techniques, aesthetic design, language expression, degree of interest, originality of work, level of endeavor, and holistic organization ability. Some of the best research products were submitted for Students' Science

Research Award and Science Composition Award. In 1994, one group's research was awarded the Kanagawa Prefectural Science Center Director's Prize for their excellent geological research on the Miura Peninsula.

In addition to organizing five field projects, science teachers designed the science lesson to be more interesting to students by making use of local rocks as well as the standard rock specimens in the laboratory. Furthermore, they also developed many special teaching materials such as a model seismograph made with short sticks and a rubber band, edible jelly layer, volcano eruption and three dimensional epicenter transparency maps.

Students seemed to enjoy this curriculum along with the science teachers, because they had many local field experiences, hands-on activities related to their life and linked to subjects like English, Japanese, social study and technology. Also interesting were the lessons that used the teacher's original teaching materials and the focus on their local environment.

4.2. Characteristics of the Five Student Field Projects

First fieldwork project: Individual student identified fieldwork to investigate geological aspects in the schoolyard:
Over a period of two months, students observed and recorded their findings about the geology of two sites. They tried interpreting their findings with their teacher's assistance and advice through the science lesson.
Second fieldwork project: Collaborative fieldwork to investigate the local extent of the rock layers and an active fault:
Students investigated the rock layers and active fault at two sites. They tried understanding them in three dimensions by comparing the geological structures they observed.
Third fieldwork project: Individual or collaborative fieldwork to collect minerals from the layers and make up the mineral specimens:
Students went to the site where they could collect the Tertiary and Quaternary volcanic rocks near school. They collected them, identified the rocks and minerals and prepared specimens that were exhibited at the school cultural festival.
Forth fieldwork project: Half-day field geological excursion to the Kenzaki area:
Teachers organized the fieldwork for students in the afternoon on Saturday. About half of the students attended the optional fieldwork in their own school district. They observed almost all geological aspects they had learned in their science class. They also investigated their local geological aspects and made some observations. They enjoyed applying their knowledge to the local environment.
Fifth fieldwork project: Integrated fieldwork during the summer vacation as a basis for the paper on local geology (individual or collaborative):
This was the last project in which students applied their knowledge and skills on geology to their own locality--the final stage of their learning of geology. Students were assigned homework for the summer vacation. They were given the guidebook on the local geology and instructed to investigate their local environment alone or in a group. They submitted their reports following the summer vacation. They were evaluated by teachers using authentic assessment techniques. Many reports were

excellent. Science teachers could evaluate their curriculum and method of their science class by assessing students' knowledge, skills and scientific thinking and interest through the reports.

4.3. Relationships with Other Subjects

Japanese (Language): Students read the science book *Ken's adventure in the wonderland of Miura Peninsula* that helped them understand how interesting the geology was and how to inquire into geological phenomena. This book was composed of about 100 pages with colored illustrations. This book was also a good example of collaborative work because science teachers wrote it with art by a fine arts teacher that was colored in by two students. Its content was reviewed for accuracy by the geology curators of the local museum, by university professors and by a teacher of Japanese composition. This book was a collaborative achievement among students, teachers and out-of school specialists. It was awarded The Science Education Prize of the Tore Science Foundation in 1994. Students read it just before the fifth field project and reacted to it in an essay. Both science teachers and Japanese teachers evaluated their compositions. Furthermore, in the second semester, students studied an essay in the Japanese language textbook on plant and animal food webs. After that, they went outdoors and investigated them with their science teachers. A science teacher as well as a Japanese teacher guided this lesson.

English: The ninth grade students had studied English for more than two years. They know many English words and like to make use of English to communicate with one another. A science teacher cooperated with an English teacher so that students could learn science through English. An English teacher, over a period of two weeks, had taught them special sentences and terminologies which would be needed to communicate in English. The ninth grade students tried communicating in English in the third field project and as many other times as possible for identifying minerals and collecting their mineral specimens.

Miura City is a sister city with the city of Warnammbool, Australia. Several students from Australia were enrolled in my lower secondary school. Several of the sister-city exchange students joined in the 9th grade science lesson in English and practiced communicating with Japanese students. The science sub-textbook *Field guide on the geology in Miura City* included the geology of sister-city Warnammbool, written in both Japanese and English. Thus, the Japanese students expanded their learning and interest on the local geology to that of a city in another country. This was one example of the internationalization of education through science.

Technology: Students learned in the science lesson that several kinds of rocks were used in constructing their school building. They also learned the technology of cutting stones and their uses in industry. For their summer assignments, some students investigated the kinds of building stones used in Miura City. Just before the winter vacation, a science teacher organized a tour to the Atomic Plants affiliated with the Rikkyo University. They were able to observe the blue light illuminating the water tank caused by the radiation from the uranium as it affects the water atoms. The president, Dr. Susumu Harasawa, held a special lecture for the students on atomic energy and its technology, and how to use it safely in society.

Social Studies: Students, during their scientific exploration in the local area, often noticed junk and garbage as well as the building stones, beach sand, minerals and the stones of historical monuments. Therefore, they expanded their interest to environmental issues. Students pointed out environmental problems in their research paper and started to consider the environment because of their experiences.

4.4. The Learning Network

Making use of public facilities and personnel: The ninth grade students found many rocks and minerals, and identified as many as possible. If necessary, teachers coordinated visits to specialists for help in identifying their samples. Through two years of experience in visiting community facilities, they learned how to get in touch with specialists and how to write a thank-you letter following their visit. Some students participated in the geological fieldwork with museum curators and helped them on their research gaining an apprenticeship experience from them. In the case of the fieldwork conducted by the Nature Conservation Society, some ninth grade students volunteered to assist the leader of the field tour. They matured from novice learners to field trip assistants and some of them will continue a role as volunteer field guide.

The museum: Some ninth grade students participated in the Sunday geological field trips guided by the curator of the Hiratsuka City Museum. Their science teacher sometimes participated in leading the museum's geological field tour, serving citizens ranging from pre-school children to the aged. Some students experienced the cooperative and collaborative learning like fossil hunting and geological investigation with curators and university professors. One day my students discovered a 10 cm fossil shark tooth from a *Caroraodon (Carucarius) Megarodon?*. It has been exhibited in the Hiratsuka City museum because it is a very important fossil. One of the students became interested in a possible career as a curator, geologist or paleontologist.

Some students also visited the Kanagawa Prefectural Museum, called The Life Planet Earth Museum, to attend the tour of the museum exhibition guided by some curators. It is well known for the state-of-the-art exhibit of the evolution of planet Earth. In addition, students could also have time to ask questions and identify their rocks. Recently this museum started establishing a special course to prepare the future geologist, instituted a geology club and developed a web site. Curators also started an outreach program to teach local rocks and fossils to students in local schools.

Science teachers also made use of the National Science Museum in Tokyo by borrowing the outreach set of samples of rocks, fossils and minerals. This collection included detailed explanations and polarized pictures of rocks so that the science lesson might be more interesting to students.

The university or institute: If necessary, teachers got in touch with geology professors from Yokohama National University and Chiba University to facilitate student's study of nature. Some students visited the atomic plant affiliated with the Rikkyo University and had a lesson from the specialist on nuclear energy. Some went to the Marine Research Laboratory affiliated with the Science and Technology Agency to see *Shinkai 6500*, a deep ocean submersible, and to learn about submarine geology and the technology for deep ocean research. These visits were opportunities for students to know about many kinds of careers in science and technology.

The nature conservation society: The ninth grade students were well prepared to investigate nature. They volunteered as assistants to the leader of the field tour. They taught young children, each other, and adults about different aspects of the natural environment. The educational effect of learning through teaching is considerable.

Development of teaching materials: Six sets of teacher materials were developed for the ninth grade curriculum. The included:

1. *Field guide on the geology in Miura City*
2. *Ken's adventure in the wonderland of Miura Peninsula*
3. Demonstration seismograph
4. Edible jelly layer
5. Erupting volcano
6. Three dimensional epicenter transparency map

5. EDUCATIONAL EFFECT

I investigated the educational effect by collecting the evaluations and opinions from students through a questionnaire. The questions asked students to evaluate the curriculum content and the learning methodologies as compared to other science classes they had taken. I used a four-scale system from "very good" to "very bad" (very good, a little better, a little worse and very bad). About 75% students among 190 at the seventh grade and 85% students among 118 at the ninth grade evaluated the overall programs as "very good" or "a little better." I also collected information and opinions about the programs from additional free responses written on the questionnaires.

From the student responses, discussions with other teachers, and my own observations, I found that there were many educational benefits of the field-centered, student-centered, inquiry-based and integrated science-centered curriculum. Some of them are:

- Students became much more interested in their local environment and appreciated their birthplace.
- Integrated learning could aid students in understanding the relationships among different school subjects in studying real situations.

- More students became interested in science and the science lesson.
- Students and teachers cooperatively held a special exhibition in the city hall in order to show and exhibit their research results and other works related to fine arts and the Japanese language. This exhibition was an opportunity for students and local people to communicate through their work. It contributed to civic understanding of school education. Parents also were proud of their children's work and understood school education much better.
- The Science Lab became a school museum since students' work was exhibited along with many teaching materials and tools.
- Some students were awarded the Governor's Prize of the Kanagawa Prefecture for their research. Many seventh grade students joined the science club because they were deeply impressed by the achievement of science club students and the fact that they were interviewed on TV for their achievement.
- Science teachers established a school-centered network.
- The established network among subjects improved the cooperative relationships among teachers.

6. SUMMARY AND CONCLUSION

Japan will introduce the *New Course of Study* (The Japanese National Curriculum) starting in 2002. The main themes of the *New Course of Study* are:

- Introduction of integrated learning,
- Five-day school system, and
- Inquiry-based learning.

The objectives of the *New Course of Study* are to foster self-learning and self-thinking ability for life-long learning. Schools are to offer a more open and relaxed atmosphere for students and teachers so that a more liberal and useful education can be planned and accomplished. The future curricula must be constructed so that teachers can respond more directly to student interests, abilities and needs. Students are to acquire the basic knowledge and skills in school curricula designed to promote education for individuality. School systems must introduce and develop evaluation standards for judging the various abilities of children in addition to simply measuring the quantity of their knowledge.

The main characteristics of the *New Course of Curriculum on Science at Lower Secondary School Level* are:

- discovery-based and inquiry-based learning,
- science related with our daily life, and
- fostering an integrated view of nature.

These characteristics are more likely to be achieved through fieldwork-centered science learning as described above in the two curriculum examples.

In order to promote the decentralization of education in the new curriculum, a new course called *The Integrated Study* and flexibility of teaching time will be introduced. Each school can organize its own curriculum based on its characteristics and ingenuity. The design and development of a school's curriculum is up to its teachers and principal.

In introducing the five-day school week system, children will have less time to learn science (30% reduction of science in the schedule). However, science teachers are given the responsibility to prepare children in the necessary basics to function as a contributing member of the nation in the global age and information-oriented society. Since inquiry-based, fieldwork-centered and student-centered science curricula are more interesting and attractive to students such would seem to be an effective way to achieve these goals for science instruction. However, teachers need more time to teach science through it than through the traditional discipline-based and teacher-centered curriculum. We science teachers must be able to create as much time as possible to organize effective science education. We can make use of the time assigned to *Integrated Study* as well as using the elective science courses to accomplish our purposes.

In upper secondary schools, additional elective courses are expected to be introduced to respond to students' interests and curiosity. Such courses, effectively taught, can deepen their learning, enhance their enjoyment of learning, and help the learner to learn in the future.

The ability and competency of teachers will be requested much more in the future. The role of teachers is not only teaching children but also coordinating, organizing and facilitating their learning and activity. Therefore, the pre-service and in-service teachers' education is much more important than before. The teacher education program, which can enable science teachers to take and guide their students outdoors, should be developed and established in the future. In the case of the Minami-shaitaura lower secondary school, science teachers in the Miura City developed their own science textbook *Field guide on the geology in Miura City* and the Miura Municipal Board of Education supported their innovative activity to develop the textbook fit for local nature and delivered it free to all the 9th grade students. The Board of Education allocated a budget of more than US$ 200,000 for the science textbook. This type of support of teacher-initiated activity is very important in encouraging creative teachers to develop locally useful materials. The science teacher leader organized an in-service teacher education program to prepare science teachers to guide their students in geological fieldwork. All science teachers in Miura City made use of the book in their science lessons. So to develop excellent local education programs, not only school staffs (teachers and principals) and the administrative staffs but also local citizens should contribute their efforts. Such cooperative and collaborative efforts will improve local education.

The philosophy of Global Science Literacy and Earth Systems Education can provide a rationale and basis for integration of not only content from the various sciences (physics, chemistry, biology and geology) but also such different subjects as fine art and social studies. Science teachers can organize integrated science learning through fieldwork-centered curriculum. Later they can also organize their curriculum expansively with science in its center involving contributions from other subjects, and by cooperating and collaborating with their teachers. Global Science Literacy and Earth Systems Education are very effective as an organizing philosophy for the total curriculum as well as for the science curriculum.

END NOTE

Definition of Expansive and Organic Learning

A.   **Expansive learning** is the expansion of learning content

(1)   Expand students' knowledge and skills and their learning from science to other subjects (expansion of knowledge and subjects)
(2)   Expand their study from their own local environment to the global-scale (expansion of content of science learning)
(3)   Expand their knowledge and skills acquired from direct experiences to abstract knowledge and thinking (expansion of thinking)

B.   **Expansive learning** is the expansion of learning methods
(1)   Change students' passive learning to active learning, for example, by asking many questions and investigating their research spontaneously (expansion of learning method)
(2)   Develop open-ended leaning (expansion of learning style)
(3)   Make use of not only the notebook but also such apparatuses as a camera, video and personal computer (expansion of learning materials)
(4)   Expand their learning field from school to museum and library, etc. (expansion of learning facilities)

C.   **Organic learning** establishes the flexible networks of knowledge and skills, content, methods
(1)   Establish the network (relationship) among their knowledge (network of knowledge and among discipline-based subjects)
(2)   Structure their knowledge hierarchically (hierarchical network)
(3)   Relate macroscopic as well as microscopic view and thinking (network of thinking, multi-lateral view and thinking)
(4)   Develop deductive as well as inductive learning flexibly (variety of learning methods)
(5)   Develop system as well as deducted approach flexibly (variety of learning approaches)
(6)   Deepen their study through communication with others among group members, instead of individual (network of communication)

Expansive and Organic Learning is developed naturally by students when science teachers guide students in enquiry-based learning through real experiences in local nature.

CHAPTER 15: THE POTENTIAL ROLE FOR GLOBAL SCIENCE LITERACY IN JAPANESE SECONDARY SCHOOLS

Victor J. Mayer, The Ohio State University, USA
Hiroshi Shimono and Masakazu Goto, National Center for
Educational Policy Research, JAPAN
Yoshisuke Kumano, Shizuoka University, JAPAN

1. INTRODUCTION

Certain events have occurred to cause the Japanese to question aspects of their educational system, especially in science. Among these were the collapse of the Japanese "bubble economy", continuing problems with the economy, and the difficulty the Japanese science and engineering communities had coping with the occurrence and after effects of the Great Hanshin earthquake in 1995. Environmental and social problems continue to increase. Public attitudes about these problems often seem uninformed. Political and economic policies toward environmental improvement have been difficult to develop and implement.

Current science education programs in Japan lack the emphasis on the Earth systems sciences of ecology and the Earth sciences that would help the public, government officials and the business community to understand the nature and cause of environmental and health problems brought on by unparalleled technological change. A survey conducted by The Ministry of Education, Science and Culture (Monbusho) revealed a number of problems and concerns with educational programs, especially in science.

1.1. Challenges in Japanese education

During the late 1990s, Monbusho was in the process of developing and implementing recommendations for revising aspects of the K-12 educational program. Its document entitled *Japanese Government Policies in Education, Science and Culture* (Monbusho, 1994) draws attention to the following set of problems in primary and secondary education:

> These include excessive competition in entrance examinations, bullying, refusal to attend school, and insufficient experience of activities in a natural environment and in everyday life. In addition, Japan is currently experiencing a variety of social changes, including the aging of the population, the shift to an information-oriented society, and internationalization. There is a strong pressure for the development of primary and secondary education enabling children to cope with these social changes appropriately. (Monbusho, 1994, p. 3)

V.J. Mayer (ed.), Global Science Literacy, 217–238.

The report especially stressed the need to revise secondary education to ameliorate these problems. Surveys of parents and school children supported the suggested changes. Their results indicated that children became less satisfied with "School Life" from primary school where 8.8% were somewhat dissatisfied or dissatisfied, to the upper secondary school were 36.8% reported these feelings. When asked what they regard as enjoyable aspects of school life, "study of school subjects" was listed by 29.0% of primary school pupils, but only 7.7% of upper secondary school students. When asked what dissatisfied them most with school life, 40.2% of upper secondary school students reported "content and process of teaching and learning". The concern regarding grades already high in lower secondary school (49.7%) increased to 53.1% in upper secondary school indicating a very high level of concern about the university entrance examinations. Further evidence was the increasing number of children attending the privately operated *jukus*. (Monbusho, 1994, pp. 6-7).

> This upward trend is attributed to the problem of excessive competition in entrance examinations, which has clearly become an issue requiring serious action in primary and secondary education in Japan. (Monbusho, 1994, p. 7)

Upper secondary schools acquire a reputation based on their students' performances on the examinations. This will influence student and parental choice of school. Heavy reliance is also placed on this score when providing children with career counseling concerning post secondary opportunities. It is these factors that lead to such excessive competition on the examinations and the rising enrollment in *jukus* (Monbusho, 1994, pp. 7-8).

The survey also revealed concerns of parents regarding experiences their children failed to have in school. When they were asked what experiences modern children lacked, the second most cited deficiency was "activities involving contact with nature". This was cited by 55.5% of the parents. The survey also sought responses on what qualities parents most wanted instilled in their children. Ranking seventh on the survey at 52.4% was a "love of nature" (Monbusho, 1994, pp. 10-11).

The report suggested that a solution to these problems could be achieved in part with the implementation of improved courses of study that are consistent with several policies, including; "Develop a respect for culture and traditions and promote international understanding" (Monbusho, 1994, pp. 17-18). Monbusho further recommended that the content of school education should be improved by expanding the range of elective subjects in lower and upper secondary schools. In addition, courses need to be based on a concept of scholastic ability that will encourage the development of skills enabling students to identify and solve problems, to think independently, and to make judgments. In so doing, educators must:

> ... change the fundamental direction of school education, which currently tends toward the delivery of large quantities of knowledge and skills as a one-way process from teacher to student. The new approach instead emphasizes the development of abilities that enable children to think and judge independently and act for themselves. (Monbusho, 1994, p. 18)

Monbusho also wanted to see an increased emphasis upon environmental education and implemented several measures to achieve this, including providing teacher reference materials and conducting seminars for teachers in charge of environmental education. The Ministry was concerned also about the "drift away from science and technology" among school children. It seeks the improvement of science education programs to ensure an adequate level of science literacy among the Japanese population. Its aim is

... to give increased priority to contact with nature, observation and experimentation, to foster problem-solving skills, and to develop scientific perception and thinking, as well as interest in and concern about nature. (Monbusho, 1994, pp. 92-3)

The greatest changes recommended by Monbusho from the previous patterns of science course content are at the upper secondary school level. In 1994, the number of courses recommended increased from nine to thirteen. This change has effectively allowed for two or three different "tracks" in science for upper secondary school students. The new courses have a strong science literacy orientation rather than the traditional "university prep" orientation to content basic to the then existing courses. The introduction of these courses was an effort to stem the growing dislike of science experienced by many students.

It is clear from the Monbusho report (1994), from discussions with many science educators in Japan, and from the accumulating evidence of the relative strengths of the economic and environmental policies of the USA and Japan, that there needs to be a new look at both systems of science education for the positive contributions they can make to each other. In this research project we take a look at a philosophy of science education, Global Science Literacy that is based in part on the National Science Education Standards (NRC, 1996) in the United States, for its relevance in solving the problems in science education and secondary education programs in Japan that were identified by Monbusho.

1.2. Preparation for study

Information, preliminary to the design of this research project, was obtained in May through November of 1996 by interviewing teachers and administrators of several Japanese secondary schools in three regions of Japan and several university faculty in an attempt to assess the impact of the recommendations of Monbusho on secondary school science offerings. These interviews revealed that teachers were unlikely to be implementing the new courses, especially in the "best" schools. They and their administrators felt that the new courses would not adequately serve their students in preparing them for the university entrance examinations. In addition, few teachers provided field experiences for their students and none seemed concerned about "internationalizing" their curricula. It was apparent in these informal discussions that the primary emphasis in science education was still on physics and chemistry as those subjects were most likely to be tested on university entrance examinations.

2. GLOBAL SCIENCE LITERACY

A program developed at The Ohio State University and the University of Northern Colorado, Earth Systems Education (ESE), provides a rationale for an international pre-college science curriculum and a global definition of science literacy. It does this by using the Earth System as the organizing theme for the development of integrated science curricula at the elementary through upper secondary school levels and the philosophy and recommended content of the National Science Education Standards (NRC, 1996). Thus using the Earth System as the organizational focus can provide science with a crucial role among other curricular subjects in helping students achieve a global understanding and perspective. In Chapter One, we have taken ESE and, with the addition of some social studies curriculum objectives, developed it into a curriculum concept we call Global Science Literacy (GSL).

Would such a curriculum concept, implemented in Japanese upper secondary schools, solve some of the problems identified by Monbusho? Would such a curriculum concept be acceptable to the science teachers expected to implement the concept in their classes? These are questions we sought to answer with a project supported by the Fulbright Senior Research program implemented through Shizuoka University and the National Center for Educational Policy Research of Monbusho.

3. CONDUCTING THE RESEARCH PROJECT

3.1. Defining the research problem

In the research reported here, the investigators interviewed science educators and teachers in Japan, to assess the relevance of ESE and GSL for Japanese secondary school science education. How can programs based on this philosophy and content address the serious issues of Japanese education cited in the Monbusho report (1994)? Specifically the problems of:

1. Internationalization of education
2. Effective environmental education
3. Improvement of attitudes toward science and technology (and hopefully, school)

In addition, the research addressed questions of barriers to implementation of revised curricula and educational practice based on Earth Systems Education and Global Science Literacy:

1. The entrance examination
2. Teacher and administrator receptiveness

3.   Social climate in schools and classroom

3.2. The research methodology.

Qualitative methods of research were used, primarily interviews of secondary school science teachers. The interviews were designed to ascertain teacher opinions concerning the questions above, as they will be the primary enforcers of any change in curriculum and teaching. The research was conducted during three months in the later part of 1998.

1.   A model of a GSL unit on typhoons (see Appendix A) was developed through a brainstorming technique. This was the concrete example used throughout the study. This unit, along with information on the research project, both in Japanese, was sent to each school at least one week before the scheduled interview session.
2.   An interview schedule (see Appendix B) was developed based upon the recommendations of Monbusho for the improvement of science education. A questionnaire was also developed based on the interview schedule to be used in large group settings. Both were translated into Japanese.

The research was conducted in two different regions of Japan--Shizuoka Prefecture and the Kanto region, which includes Tokyo and its vicinity.

3.3. The Shizuoka Component

Shizuoka is a large city located about 100 kilometers southwest of Tokyo. The first stage of the research in this region consisted of a full day teacher-training program on October 24, 1998 at Shizuoka University. It started with a lecture on the nature and rationale of Global Science Literacy followed by several cooperative learning experiences in which the teachers brain-stormed ideas for developing GSL curricula. The workshop concluded with an illustrated lecture on Earth systems aesthetics. One of the major goals of the in-service training was to introduce the concept of Global Science Literacy applied to the upper secondary school science curriculum. Subsequently we conducted interviews of teachers in four different upper secondary schools in Shizuoka Prefecture. At least one science teacher in each of the schools had attended the in-service training.

3.3.1. Description of data collection methods
1.   A questionnaire was developed based upon the interview schedule in Appendix B. Part 1 of the questionnaire was administered at the beginning of the workshop. It assessed the extent to which teachers were attempting to accomplish certain goals stated by Monbusho in the 1994 report. Part 2 was administered at the end of the workshop. It assessed teacher opinions of the extent to which GSL could help achieve those goals of Monbusho for the science curriculum.

2. Interviews were conducted in each of the four upper secondary schools about two months after the in-service training workshop. This was a group interview in which all of the cooperating teachers participated at the close of classes. Each was questioned individually, following the interview schedule (Appendix B). Each session was tape-recorded. The transcripts of the tapes were then translated and analyzed. We intended to compare the opinions of teachers who attended the daylong workshop with those who were only given a four-page description of GSL (Appendix A) immediately before the interviews.

Table 1. Schedule of in-service training and upper secondary school teacher interviews

Activity	Date	Location	# Teachers
In-service Training	Oct. 24	University Campus	(10)*
Interview 1	Nov 20	School A	2(1)
Interview 2	Dec 4	School B	4(1)
Interview 3	Dec 4	School C	11(1)
Interview 4	Dec 7	School D	
Total			32(5)

* Parentheses indicate the number of teachers who participated in the In-service training program.

We anticipated that there would be significant differences on opinions concerning GSL and its utility in reaching Monbusho recommendations for reform between the teachers who attended the in-service GSL training and the teachers who did not attend, since these teachers would have had a more prolonged and complete introduction to the concepts and rationale behind GSL.

3.3.2. Data presentation and analysis
Workshop data

1. Questionnaire: A four option system for the responses was used for the pre-questionnaire, from a "must include" option for response to a "don't include" option as an important objective of "your teaching". A lower score thus shows higher valuing of the objective by the teacher. For the post-questionnaire, a five option system of responses was used, from GSL promising to be "much more effective" to "not at all effective" in accomplishing the objective. A smaller score indicates greater support for GSL as being effective in accomplishing the stated objectives. In order to compare the two scores, the four-point scale scores were converted to the five-point score. The adjusted scores for each item on the pre-questionnaire and the raw scores for the post-questionnaire are shown in Table 2. There were statistically significant differences between the total adjusted pre-questionnaire scores and the total post-questionnaire scores using

the T-test (p=0.032).

2. Conclusions from Questionnaire: It would appear therefore, that these teachers are supportive of the objectives stated on the questionnaire indicating that all should be included in science curricula. They are also of the opinion that GSL, if implemented in their schools, would be more effective than their current curriculum in achieving these objectives.

Table 2. Adjusted scores for the pre questionnaire and raw score for post-questionnaire

Item Number	Pre-Questionnaire Scale 4 to 1 N=10 (adjusted to 5)	Post-Questionnaire Scale 5 to 1 N=10
1.Understanding environmental issues	1.60	1.30
2. Learning science behind the issues	2.11	2.00
3. Value environmentally friendly technologies	1.90	1.80
4. Internationalization	2.10	1.56
5. Developing interest in science	1.30	1.40
6. Developing life-long learning abilities	1.90	1.56
T-test Result; p=0.032 Totals	10.91	9.62

3. Teacher comments: Written comments on the problems or barriers for implementing GSL in Japanese upper secondary schools were contributed by ten of the teachers. They are summarized in Table 3 where a letter for each teacher having a given comment is included in the first column. The letter indicates the teaching responsibility of that teacher.

Table 3. Written comments by teachers on post questionnaire

Number of Teachers (by teaching area)	Comment
P, E, P, LS ,B, B	Entrance examination
E, LS, B	Difficulty in changing science teachers' attitudes

P	Introduce in "project research"
C	Approach can be included in my classes
E, P	Science education policy must change
P	Difficulty of learning
LS	Must integrate or we will fail to solve Earth's problems
E	Teachers do not have time to learn new things
LS, E	Narrow preparation of science teachers

P=Physics, C=Chemistry, B=Biology, E=Earth Science, LS=Lower Secondary

4. Summary of teacher comments: It is clear from the teachers comments that the university entrance examination is perceived as the major barrier in implementing change, even by the lower secondary school teacher. The difficulty in changing teacher attitudes was seen by the teachers as another factor, probably resulting from their narrow preparation in a single science discipline.

5. Conclusions from comments: Global Science Literacy was an entirely new term and concept for most if not all of the teachers. It would be a difficult concept to communicate to teachers over a long period in an intensive workshop, much less a one-day program. It is especially hard for secondary school science teachers to understand the philosophy and content of GSL in Japan. However, it was apparent that most teachers accepted the idea of GSL, as they understood it, as being an appropriate basis for curriculum revision. They saw such revision unlikely to be accomplished however, because of the problems of university entrance examinations and lack of preparation of teachers.

Results from Interviews

Four upper secondary schools were visited in this stage of the project. The science teachers in each of these schools met in a group, but were interviewed individually so that all science teachers heard each other's responses. Instead of strictly following the interview schedule (Appendix B) the interviews were less structured and open than those conducted in the Kanto region study. Sixty-seven specific comments resulted from the interviews (Table 5). They fell into five general areas.

1. Opinions in support of GSL as a basis for the development of science curriculum in Japanese upper secondary schools:
   There are twelve ideas expressed which support the likely effectiveness of GSL

as a foundation for the science curriculum. Some of the ideas focused on the importance of the nature of science, not currently a focus in Japanese science education. In addition, some teachers believe that science can be learned more effectively through a systems science course such as GSL. Two of the teachers who had also participated in the in-service training were very strongly supportive of GSL.

2. Opinions not supportive of GSL for Japanese schools:
   There are twelve comments not supportive of implementing GSL, however; only one teacher believed his traditional way of teaching science was the best way. Some teachers think that a GSL course would need more time allocated to it in the school day because of the complex and detailed subject matter. In addition, GSL teachers would need to be prepared in a much different way than is current. Some teachers think that to offer GSL courses in the upper secondary school level might lessen the quality of science education. In their opinion, this would be because students learn higher systems science ideas without understanding basic ideas in science.

3. Problems of implementation caused by university entrance examinations:
   Many teachers pointed out the influence of the university entrance examinations on their teaching and science curricula. Only if there were content reflecting GSL in the entrance examinations would science teachers in upper secondary school be able to promote the GSL programs.

4. Opinions regarding the need for administrative and teacher education support:
   Teachers need to be prepared for implementing GSL courses. In addition, there would need to be well-organized teacher education programs for GSL. In general, school facilities and equipment would have to be improved including the provision of good libraries and Internet access. There would also need to be strong administrative policy and support for systems science curricula.

5. Suggestions for the implementation of GSL:
   Some teachers did not believe that GSL curricula would adequately cover all fields of science. They identify GSL as an Earth science based notion of scientific literacy. For the implementation of GSL, it is important to develop programs that are relevant within students' contexts, have connections to students, and do not cause too much anxiety about our future. In addition, mathematics needs to be incorporated into science curricula at least to the same level of present programs in Japan.

4.3.3. Conclusions from the Shizuoka Study

In this study, it is clear that few science teachers in Shizuoka are ready to develop GSL in their schools for many reasons. However, it is also true that many science teachers understand the desirability of such a curriculum for their schools and would be willing to teach it if they had the proper training and support.

The *New National Standard Curriculum, Course of Study* for elementary and lower secondary schools was officially announced in December 1998. The upper secondary school course of study was announced in March 1999. They incorporate a major change in science. From the elementary school to upper secondary school the Course of Study includes a sequence called "Integrated Time", which focuses on environmental issues, international understanding, information processing and well-being. In the context for developing a life-long learning society, "an Integrated Time" will be introduced in every school from elementary through lower secondary school. Organization of "an Integrated Time" will depend on each school's situation. So "an Integrated Time" so-called *Sougou-teki Gakusyu* in Japanese could be developed around the schema of GSL.

In addition, for the upper secondary school level there will be three new subjects in science including Basic Science, Integrated Science A, and Integrated Science B. The focus of Basic Science will be on the nature of science. Students will try to learn how science evolved using historical case studies in which teachers will try to use topics close to students' lives. Within the Integrated Science A, the focus will be the principle ideas in Physics and Chemistry. In Integrated Science B, the focus will be the principle ideas in Biology and Earth Science. All upper secondary school students have to choose one of the three.

3.4. The Kanto Component

This component consisted of interviews of teachers in two different upper secondary schools and one prefecture's educational center all located in the Kanto region in and around Tokyo. The same documents were used as in the Shizuoka study. In addition, the questionnaire was completed by about 50 teachers enrolled in a teacher workshop in Gifu Prefecture conducted by the Earth science education research component of the National Center for Educational Policy Research of Monbusho.

3.4.1. Data collection procedures
1. An oral presentation on GSL was given in English with translation into Japanese to the teachers in Gifu, none of whom were involved in the school interviews. The presentation was followed by completion of the questionnaire, slightly modified in format to make it appropriate for this use. Usable questionnaires were obtained from 33 of the teachers.
2. The GSL model (Appendix A) was distributed to the science teachers in each of four schools in the Kanto region. This was followed within the same week by interviews held in each of the schools. All of the teachers met following the end of classes for the day. Each teacher was asked to respond to each of the interview questions in a set sequence. The interview schedule (Appendix B) was followed.

The major question to be answered concerned teachers ability to understand GSL and its implications for reaching Monbusho objectives. We also were interested in determining the relative effectiveness of paper descriptions of GSL followed by interviews and that of a short oral presentation followed by a paper and pencil questionnaire for securing data.

3.4.2. Data from the oral presentation at Gifu

Part one of the questionnaire included questions on the six objectives stated by Monbusho for future school programs. Teachers indicated their opinions of the importance of their including each of the objectives in their teaching on a four option system from "must include" to "don't include". Responses to the option of "must include" was scored as a four, down to one for "don't include". Responses would be an indication of the degree to which they already include the objective and therefore the importance that they themselves place on the objective.

Table 4. Responses to questionnaire items.

Item	Part 1: Includes Objectives (4 pt max)	Part 2: Advantages with GSL (4 pt max)
1.Understanding environmental issues	3.67	3.03
2. Learning the science behind issues	3.22	3.20
3. Value environmentally friendly technologies	3.03	3.15
4. Internationalization	3.18	2.91
5. Developing interest in science	3.64	3.39
6. Developing life-long learning abilities	3.21	3.14
Totals	19.95	18.82

Part two of the questionnaire included the same six objectives. Teachers gave their opinions of the probable effectiveness of GSL for reaching the objectives on a five option system, from GSL promising to be "much more effective" to "not at all effective" in accomplishing the objective. A score of five was ascribed to "much more effective" and one to "not at all effective" than their current practices for reaching Monbusho objectives. Part 2 scores were converted to a four-point basis (see Table 4) to take account of the different scale (five points to four points possible). Higher scores on both scales are interpreted as more positive. Note that the valuing of the Shizuoka scale is the reverse of this one. Also the conversions of the scores on each of the two questionnaire parts is the reverse with this data converted to a four-point basis and the Shizuoka data converted to a five-point basis. Two different research teams performed the analysis accounting for this difference. Unfortunately, this will lead to some confusion in the analysis of the data.

A total neutral response with the Shizuoka teachers would have been 18. For the Gifu teachers it was 15. On Part 1 Shizuoka, the mean response was 10.91, Gifu, 19.95 (Tables 2 and 4). On Part 2 Shizuoka, 9.62 and Gifu, 18.82. The departure from neutral, though positive for the Gifu teachers, was much less on both scales than for the Shizuoka teachers. Thus, these teachers were not at all as positive, either about including Monbusho objectives in their curricula or about the potential of GSL for addressing these objectives, as were those teachers who completed the Shizuoka University GSL workshop. They were also slightly less positive about the potential of GSL for addressing these objectives than they seemed to be about including the objectives in their current curricula.

Some 29 teachers made comments on their questionnaires regarding what they saw as problems in implementing curricula with the objectives of GSL. They are summarized in Table 5 below.

Table 5. Summary of teacher comments.

Comments (from 29 teachers)	Number of Teachers
Teacher's knowledge and in service training	12
University entrance examinations	11
Time available to teach science	9
Teaching materials	4
Coordination with other subjects	3
Number of science teachers	2
Evaluation	2

These results are consistent with those cited on the Shizuoka University workshop evaluation. Teachers see university entrance examinations as being a major impediment to change, along with their own knowledge and consequent need for in-service training. Also high on the problem list is the time they have available to teach science. In their minds, it is insufficient to teach courses based on the GSL philosophy.

3.4.3. Summary of interview responses of the Kanto area teachers: Part One

Regarding the potential for GSL in providing effective Environmental Education in Japan: Twelve teachers among fifteen agreed with the effectiveness of GSL in environmental education in Japan. Only one biology teacher disagreed with it because he felt that GSL deals with more geological content than biological. He felt that there needed to be more biological content included in order to result in effective environmental education. However, overall, GSL was regarded as an effective approach by science teachers in accomplishing the objectives of environmental education.

Regarding the potential of GSL developing an understanding for the use of sustainable technology: Eight teachers among nine have positive responses to the potential of GSL for sustainable technology. Almost all teachers appreciate the potential of GSL for developing an understanding for the use of sustainable technology in our society. Some teachers did not adequately understand the meaning of sustainable technology in science education. GSL gave them an opportunity to know about it.

Regarding the potential of GSL for developing international understandings about science and society: Twelve teachers out of fifteen think that GSL courses could be effective in developing international understandings. None of the teachers disagreed, because the GSL approach deals with global concepts and makes use of computers and the Internet to collect information

Regarding the potential of GSL for improving students' attitudes and interest in science: Twelve teachers out of fifteen think GSL could be effective for improving students' attitudes and interest in science. Only one teacher disagreed. This potential is there because GSL curricula would deal with things familiar to students and encourage students to engage in science related activities beyond the boundary of the traditional science disciplines. Some teachers indicate an integrated approach such as GSL should be taught to students in lower secondary schools before it is introduced to upper secondary schools.

3.4.4. Summary of interview responses of Kanto area teachers: Part Two
Regarding the potential of GSL as an introductory science course in upper secondary education: Eight teachers out of sixteen recommend GSL as an introductory science course in upper secondary school. Many teachers recommended GSL as one of several elective courses. The most important problem and barrier is limited teaching time since they believed that a GSL course would need more time to teach.

Regarding the potential of GSL as the organizing principle for all science courses in upper secondary education: Eight teachers out of sixteen recommended GSL as the organizing principle for all science courses in upper secondary education. Japanese upper secondary school teachers get accustomed to teaching courses confined narrowly to disciplinary fields like physics, chemistry, biology and Earth science. However, they are worried about whether present science teachers trained in those narrow fields can adequately respond to the needs of teaching across several disciplines as would be necessary in GSL organized courses.

3.4.5. Summary of the Kanto area interviews

Almost all teachers think GSL is an effective way to achieve the objectives of environmental education, including understanding sustainable technologies, achieving an international perspective and improving student's attitudes and interest in science. But only half of the teachers want to change science courses from the present discipline-based course to the GSL systems course. Ideally they agree with the GSL approach but really they don't want to change the present discipline-based subject which they have become accustomed to teaching. There is one reason why they believe that at present the discipline-based teaching is more effective then a GSL course and that is for preparing students to get a better grade in the university entrance examinations. They are always concerned about the university entrance examination and think that teachers should teach students discipline-based subjects in order to pass it. Though GSL courses could be effective and attractive to students, the teachers were concerned about their abilities to understand and teach content from several different fields of science and the lack of time in the current science schedules to teach such a course. GSL content was seen as potentially more interesting to students but seen to require much more preparation and effort on the part of the teachers. However, they feel that some students would love such an integrated approach as GSL. Therefore, teachers recommend GSL as an elective course in upper secondary school at present. They see the GSL concept as good, effective and promising but implementing GSL teaching would be difficult in Japan at present.

4. GENERAL CONCLUSIONS FROM THE STUDY

4.1. Research designs used in the study

Of necessity, several different approaches were used in this study to obtain upper secondary school science teacher opinions of their efforts at implementing certain objectives of Monbusho for education and of the feasibility of using a Global Science Literacy curricular approach in upper secondary school science curricula.

1. Comparison of teachers' knowledge and attitudes regarding GSL as a result of their participation or non-participation in the Shizuoka University workshop: Apparently only two of the ten teachers who had participated in the workshop expressed more positive attitudes toward GSL during the subsequent interviews than did their colleagues who had not participated in the workshop. Not enough information, however, is available to project any conclusions from this comparison.

2. Open versus structured format of interviews: The Shizuoka interviews followed the interview schedule, as did the Kanto region interviews. However, in Shizuoka, teachers were allowed to answer questions and discuss each other's responses, whereas in the Kanto interviews teachers were questioned in a rigid sequence. No discussion occurred until the end of the interviews. The Shizuoka questioning was somewhat more informal and unstructured. It seems that the Kanto interviews resulted in more diverse information. This may be the result of

the differences in interview techniques, personalities of the interviewers, characteristics of the teachers, or a combination of factors.

3. Comparison of questionnaire responses following a full-day workshop on GSL and a 45-minute long oral presentation: Teacher's responses on the questionnaire following the full-day workshop were consistently more positive, concerning both the Monbusho objectives and the GSL portion of the questionnaire. This, on the surface at least, supports the need for longer workshops in acquainting teachers with both the Monbusho objectives and the nature of Global Science Literacy.

4.2. Problems encountered in implementing Global Science Literacy

In order to implement courses structured along the lines of GSL in Japan three barriers must be solved. The first barrier is related to a single teachers' ability to teach a wide integrated field of science. The Japanese upper secondary school science teachers were trained to teach their specialty, physics, chemistry, biology or Earth science. If teachers teach a GSL course at upper secondary school, they would need to have extensive in-service training and pre-service training. However, the present teacher training system is not able to respond to this need. One alternative would be for teachers to teach in a team comprised of teachers of different disciplinary specialties. This would also require extensive in-service training on team teaching procedures and the nature of the system sciences. An integrated course was introduced in the present science curriculum but it was not effective in part because it was only a mixture of discipline-based subjects (physics, chemistry, biology and Earth science). It was not conceptually and systemically organized such as a quality GSL course would be.

The second problem is the extended time necessary to teach a GSL based science course. The systemic and student-centered approaches need more time than the present discipline-based, fact oriented and teacher-centered approaches. Recently Monbusho decided to reduce science-teaching time. This trend is the opposite of what would be necessary for GSL at least as perceived by the teachers in this study. The newly established *Integrated Learning Time* might be useful for initiating GSL experiences. However, the use of *Integrated Learning Time* depends on each teacher and each school. Each teacher must develop his or her own integrated learning curriculum. There is no national curriculum for it.

The third and most difficult barrier is the system of university entrance examinations. Upper secondary school teachers are very concerned about their student's success on the entrance examinations and structure their curriculum and teaching methods to respond to it. The university entrance standard is mainly dependent on getting a better grade on difficult discipline-based problems. In order for students to pass them, science teachers concentrate their teaching on the discipline-based curricula to ensure that their students obtain the best possible grades on their entrance examinations. They don't feel this objective can be achieved with integrated or systemic curricula. For teachers to implement any type of science program that departs from the traditional discipline-based sciences of physics, chemistry and biology the university entrance examination system must be changed.

In some academically low level upper secondary schools, where few graduates will go on to university, teachers introduced integrated science courses in an effort to interest students more in science learning, This was possible since the teachers did not have to worry about the entrance exam. The time will come when the falling birth rate and thus number of students interested in university education, will allow a broader cross section of students into university education. Then, perhaps, the examination problem will disappear and integrated and systemic learning will be introduced more easily than is possible today.

5. SUGGESTIONS FOR FURTHER STUDY

In order for GSL to be used extensively in Japan, we must accomplish the following.

♦ Develop many GSL units such as the Typhoon Unit and organize them into a three-year curriculum: This would provide a concrete example of a GSL curriculum for evaluation by science teachers and administrators. If the three-year-course curriculum based on GSL is developed, Japanese upper secondary school science teachers can judge whether a GSL course will be effective for Japanese science education.

♦ Develop exemplary teaching materials and textbooks: At least one model textbook on GSL should be developed to encourage the dissemination of GSL. For example, a text using the systemic science unit on typhoons should be developed in order to demonstrate to teachers how GSL would integrate physics concepts such as force, momentum and inertia, and essential biological and chemical concepts. Additional units should be developed and incorporated into such a text. Many good teaching materials and innovative textbooks facilitate teachers' attempts to develop their own original teaching materials.

♦ Develop authentic evaluation methods for GSL curricula.

♦ Develop programs for in-service and pre-service education: A major barrier to the implementation of GSL courses is teacher background. Science teachers must be prepared for teaching systemic science courses. The promoters and facilitators of GSL should present concrete programs for the preparation of in-service teachers and pre-service students.

♦ Establish a learning network to sustain GSL learning: Student-centered, inquiry-based, systemic learning cannot be sustained and supported by the school alone but must be assisted by outside community resources and establishments such as museums, communications establishments, environmental organizations, etc. Teachers need to coordinate student learning more effectively by contacting such community learning resources and staffs. Therefore, teachers should play many roles of teacher, supporter, facilitator and coordinator. The school administrators should develop and sustain such community-based learning networks.

AKNOWLEDGEMENTS

This research was conducted with support from a Fulbright Senior Researcher Grant to Dr. Mayer, the National Center for Educational Research of Monbusho, and Shizuoka University.

REFERENCES

American Association for the Advancement of Science. (1989). *Science for All Americans*. Washington D.C.: AAAS.
Ministry of Education, Science and Culture. (1994). Author.
National Research Council, (1996). National Science Education Standards. Washington D.C.: National Academy Press.
Secretary's Commission on Achieving Necessary Skills. (1992). *Learning a Living: A Blueprint for High Performance*. Washington, D.C.: U.S. Department of Labor.

APPENDIX A

INFORMATION AND MODEL UNIT SENT TO EACH SCHOOL

GLOBAL SCIENCE LITERACY

Global Science Literacy seeks to broaden students understanding of the nature of science. Its basic thesis is that science is the process that we as humans use to understand the world we live in and its environment in space. Therefore all science instruction should start with some aspect of the Earth system (biosphere, solid earth, atmosphere and hydrosphere), the solar system, or the galaxy and universe we inhabit. Since we live on Earth, the central subject for science teaching should be the Earth. Scientists and science teachers have special methods to investigate earth processes. One of these methods you are already acquainted with from your previous study of science. You are well versed in conducting science experiments and the need to identify, isolate and control variables. This is what we call the reductionism method of science. It has been very useful in obtaining knowledge about those earth processes that lead to technical applications. There are other approaches to the science of the Earth, however, that are often more useful in studying the natural processes; how they have worked in the past to produce the characteristics of our environment and how they may work in the future. This has been called the systemic method of science. Unless you have studied ecology or the Earth sciences, you may not be familiar with this type of science. Both types of science method are represented in curricula based on GSL.

In GSL, it is important to acquaint students with the big ideas in science, whether they have been developed by the sciences of physics, chemistry, biology or the Earth sciences. All ideas that are taught however, are taught through their application in understanding some aspect of an Earth system. Thus, the physics concept of density can be learned through its role in typhoons, volcanic eruptions, or how fish maintain buoyancy. The following characterize GSL curricula:

1. .Instruction on a big idea always begins with student experiences with nature. The big idea is placed in the context of its occurrence as an aspect of an Earth system; for example, how air density and pressure functions in the origin and activity of a typhoon.
2. The aesthetic aspects of the Earth system process being studied are included in science instruction. For example, this may include observing the visual beauty of the typhoon in a satellite photograph, or the representations of typhoons in Japanese literature, art or music.
3. All concepts important for a student understanding of the world about him/her are imbedded in an Earth system context whether they have been considered a domain of physics, chemistry biology or any other science.
4. Although instruction starts with a local (the student's country) example of the Earth process, examples of this process from other countries around the world are also drawn. In this way, a Japanese student can obtain a worldwide perspective of the natural environment.
5. GSL emphasizes science as a universal medium for communication among the Earth's varied cultures. Its constant reference to data and proof being a model for communications in other realms of human endeavor such as social and cultural interactions and political activities.
6. It includes a study of the use of technology in assisting students to understand the Earth

system process being learned and the technology's application in the every day life of the student. For example, the use of satellite and computer technology in tracking and predicting the influence of typhoons would be included in a unit on typhoons.

7. GSL includes the use of system science methodologies in investigating Earth system processes as well as the reductionism methods of most current science curricula. For example, students would study data obtained from charting typhoons over a 10-year period to determine how they tend to act in the region of Japan.

Examples of GSL units of instruction:

1   Typhoons and weather.
2.   Volcanic activity and its effects on the biosphere
3.   Earthquakes and earthquake prediction
4.   Fishing and the production of sushi
5.   Water motions in the ocean and their human implications

EXPANSION OF UNIT ON TYPHOONS AND WEATHER

Instructional topics and activities:

1. Introduce the unit with haiku, art or music related to typhoons or weather. Integrate aesthetic content throughout the unit.
2. Teach the concept of differences in density of air; temperature, moisture. Include effects on air pressure, ocean surface, and wind generation. Employ both laboratory investigations and microclimate investigations in the out of doors.
3. Study the effect of solar radiation on the ocean surface and atmosphere. Use laboratory investigations and out door activities.
4. Introduce concepts of precipitation, evaporation, condensation, and energy in water vapor. Use laboratory investigations
5. Study air-ocean interface and interactions. Learn from resource materials.
6. Learn how plants and animals of the ocean and coastal areas have adapted to typhoon activity. Use resource materials and coastal field trips.
7. Learn about the interaction with land and its influence on typhoon activity. Can include a study of force, momentum, and inertia. Use field and laboratory investigations.
8. Learn about the types of instruments used in gathering data on atmospheric conditions. Use laboratory experiences and resource materials.
9. Learn how technology can be used to predict, track and ameliorate the effects of typhoons. Use of Internet data and other sources of information.
10. Study the effects of typhoons on human activities on land and sea. Use television news reports, documentaries, and personal experiences.
11. See how scientists from around the world cooperate to study tropical storm activity. Study examples of other tropical storms, monsoon rains and hurricanes, which occur in other areas of the world. Use resource materials.

Concepts and processes learned through such a unit:

Related to Earth as a planet
1. Incidence of solar radiation with latitude
2. Rotation as it induces the Coriolis effect

Related to the ocean
3.   Heat capacity of water
4.   Evaporation at the surface
5.   Influence of air pressure on sea level
6.   Generation of waves by wind
7.   Current movements

Related to the atmosphere
8.   Influence of moisture and temperature on air density
9.   Causes and variations in air pressure
10. Movement caused by differences in air pressure
11. Adiabatic cooling
12. Precipitation and the release of energy

Related to the influence of land
13. Change in elevation and rising of air mass
14. Cooling results in precipitation
15. Produces flooding and erosion.
16. Concepts of force, momentum, inertia.

Related to plant and animal life
17. Adaptations of ocean animals to storm activity.
18. Mechanisms used by coastal organisms to withstand effects of storms.

Related to technology
19. Instruments used to measure temperature, humidity, air pressure, wind velocity and direction.
20. Use of satellites in tracking typhoons.

Related to human society
21. How the art and culture of the student's country have included the subject of typhoons.
22. An appreciation of the effect of typhoons on the student's society and economy.
23. An understanding of how science can contribute to communication on common problems between different cultures.

INTERVIEWS

In a week or two, we will be meeting with you and several of your colleagues. We will be asking you certain questions regarding Global Science Literacy to seek your opinions regarding this new approach to science curriculum. The questions will focus on the following issues:

A.   Can programs based on this philosophy and content address the serious issues of Japanese education cited by Monbusho? Specifically the problems of:
1. Internationalization of education
2. Effective environmental education
3. Improvement of student attitudes toward science and technology

B.   What might be the barriers to implementing revised curricula and educational practice based on Global Science Literacy?
1. As the basis for a one year integrated science course in the 9th or 10th year?
2. As the basis for integrating the entire science curriculum in the upper secondary school?

APPENDIX B

THE INTERVIEW SCHEDULE USED IN KANTO AREA TEACHER INTERVIEWS

Teacher Interview Schedule:

1.   Which of the following, if any, are important goals for science teaching in your classes? Please comment on your opinions of each.

A.   Improve your students knowledge of environmental issues and of the science important for solving environmental problems and for developing "environment friendly" alternative technologies.

B.   Internationalization of education.

C.   Develop your student's interests in science and their positive attitudes and abilities for life long learning in science.

2.   We are suggesting that Global Science Literacy can provide the foundation for the development of "integrated" courses in science. What do you feel the potential of such a course might be in comparison with what you currently teach, for accomplishing each of the following for the typical upper secondary school student?

A.   For developing the scientific knowledge for the understanding of environmental problems and their solution and for the development of "environment friendly" alternative technologies.

B.   The international understandings about science and society.

C.   Improving student interest in science and developing positive attitudes and abilities for engaging in life long learning in science.

3.   What do you see as potential problems or barriers in the implementation of an integrated science course based on Global Science Literacy?

A.   As an introductory science course in upper secondary education?

B.   As the organizing principle for all science courses in upper secondary

Index